Science and Conservation in African Forests

Tropical forests, like the great apes and other wildlife that inhabit them, are under threat throughout their range. This unique case study from Kibale National Park in Uganda illustrates that a vital aid to forest conservation is the establishment of a long-term research facility. Biological research, of the type pioneered by legendary figures Jane Goodall and Diane Fossey, not only produces important data for the design of management practices, but also has many unanticipated benefits that promote conservation, including improved community relations, positive economic outcomes, and training of future generations. Every well-established study of great apes from Guinea and the Ivory Coast to Tanzania and Rwanda shows the same pattern of benefits. Future environmentalists can use the lessons learned in this book as a model for conservation through research.

RICHARD WRANGHAM is the Moore professor of Biological Anthropology at Harvard University and a patron of the Great Ape Survival Project. Since 1987 he has directed the Kibale Chimpanzee Project in Western Uganda, studying the behavior and ecology of a community of wild chimpanzees.

ELIZABETH ROSS has been the Director of The Kasiisi Project since 1997. The Project operates in the rural communities bordering Kibale National Park, Uganda and provides educational support to students and staff in government Primary schools. She has a B.Sc. in Biology and a Ph.D. in Immunology from Edinburgh University.

Science and Conservation in African Forests

The Benefits of Long-term Research

Edited by
RICHARD WRANGHAM
Harvard University, Massachusetts, USA

ELIZABETH ROSS
Kasiisi Project, Uganda

This book is a result of a workshop on Long-term Research and Conservation held to celebrate the twentieth anniversary of the Kibale Chimpanzee Project and the tenth anniversary of the Kasiisi Project at Makerere University Biological Field Station, Kibale National Park, Uganda, June 15–17 2007.

CAMBRIDGE UNIVERSITY PRESS
Cambridge, New York, Melbourne, Madrid, Cape Town, Singapore, São Paulo, Delhi

Cambridge University Press
The Edinburgh Building, Cambridge CB2 8RU, UK

Published in the United States of America by Cambridge University Press, New York

www.cambridge.org
Information on this title: www.cambridge.org/9780521896016

First published 2008

Printed in the United Kingdom at the University Press, Cambridge

A catalog record for this publication is available from the British Library.

Library of Congress Cataloging in Publication Data
Science and conservation in African forests : the benefits of long-term research / edited by Richard Wrangham, Elizabeth Ross.
p. cm.
Includes bibliographical references.
ISBN 978-0-521-89601-6 (hard back) – ISBN 978-0-521-72058-8 (paper back)
1. Forests and forestry–Research–Uganda–Kibale National Park. 2. Forest conservation–Uganda–Kibale National Park. I. Wrangham, Richard W., 1948- II. Ross, Elizabeth, 1950- III. Title.

SD356.54.U332K557 2008
634.9'2096–dc22 2008016135

ISBN 978-0-521-89601-6 hardback
ISBN 978-0-521-72058-8 paperback

Contents

Contributors

Fred Babweteera, Budongo Forest Project
Budongo Conservation Field Station, PO Box 362, Masindi, Uganda

Zoro Bertin Goné Bi, Taï Chimpanzee Project
Centre Suisse de Recherches Scientifiques,
01 BP 1303, Abidjan, 01 Côte d'Ivoire

Michael Binford
Department of Geography, 3141 Turlington Hall PO Box 117315,
Gainesville, FL 32611-7315, USA

Christophe Boesch, Taï Chimpanzee Project
Max-Planck Institute for Evolutionary Anthropology,
Deutscher Platz 6, 04103 Leipzig, Germany

Hedwige Boesch, Taï Chimpanzee Project
Max-Planck Institute for Evolutionary Anthropology,
Deutscher Platz 6, 04103 Leipzig, Germany

Colin Chapman, Kibale Fish and Monkey Project
Department of Anthropology and McGill School of Environment,
855 Sherbrooke St West, McGill University, Montreal, Quebec, Canada
H3A 2T7

Lauren Chapman, Kibale Fish and Monkey Project
Department of Biology, McGill University, 1245 Dr. Penfield
Montreal, Quebec, Canada H3A 1B1

Anthony Collins, Gombe Stream Research Institute
Lake Tanganyika Catchment Reforestation and Education, PO Box 1182, Kigoma, Tanzania

Katie Fawcett, Karisoke Research Centre
Dian Fossey Gorilla Fund International, Rwanda

Thomas R. Gillespie, Kibale Eco-health Project
University of Illinois, Department of Veterinary Pathobiology, 2001 South Lincoln Avenue, Urbana, IL 61802, USA

Tony Goldberg, Kibale Eco-health Project
University of Illinois, Department of Veterinary Pathobiology, 2001 South Lincoln Avenue, Urbana, IL 61802, USA

Abe Goldman
Department of Geography, 3141 Turlington Hall, PO Box 117315, Gainesville, FL 32611-7315, USA

Jane Goodall, Gombe Stream Research Center,
The Jane Goodall Institute US Headquarters, 4245 North Fairfax Drive, Suite 600 Arlington, VA 22203, USA

Joel Hartter
Department of Geography, University of New Hampshire, 127 James Hall, 56 College Road, Durham, NH 03824, USA

Ilka Herbinger, Taï Chimpanzee Project
Centre Suisse de Recherches Scientifiques, 01 BP 1303, Abidjan 01, Côte d'Ivoire

Gilbert Isabirye-Basuta
Department of Zoology, Makerere University, PO Box 7062, Kampala, Uganda

Natarajan Ishwaran, Division of Ecological and Earth Science, UNESCO
1 rue Miollis, 75732 Paris cedex 15, France

John Kasenene, Makeree University Biological Field Station
MUBFS, PO Box 409, Fort Portal, Uganda

Makan Kourouma, Bossou Chimpanzee Project
IREB, Bossou, Guinea

Florence Landsberg, World Resources Institute
10 G Street, NE (Suite 800), Washington, DC 20002, USA

Nadine Laporte, The Woods Hole Research Center
149 Woods Hole Road, Falmouth, MA 02540-1644, USA

Jeremiah S. Lwanga, Ngogo Chimpanzee Project
MUBFS, PO Box 409, Fort Portal, Uganda

Moses Mapesa, Executive Director, Uganda Wildlife Authority
Plot 7, Kira Road, Kamwokya, PO Box 3530, Kampala, Uganda

Tetsuro Matsuzawa, Bossou Chimpanzee Project
Primate Research Institute, Kyoto University Kanrin, Inuyama-City, Aichi 484-8506, Japan

Michio Nakamura, Mahale Chimpanzee Project
Human Evolution Studies, Department of Zoology, Graduate School of Science, Kyoto University, Kyoto 606-8502, Japan

Toshisada Nishida, Mahale Chimpanzee Project
Japan Monkey Centre, 26 Kanrin, Inuyama 484-0081, Japan

Emmanuelle Normand, Taï Chimpanzee Project
Max-Planck Institute for Evolutionary Anthropology, Deutscher Platz 6, 04103 Leipzig, Germany

Clive Nuttman, The Tropical Biology Association
Department of Zoology, Downing Street, Cambridge CB2 3EJ, UK

William Olupot, WCS Albertine Rift Programme
PO Box 7487, Kampala, Uganda

Patick Omeja, Kibale Fish and Monkey Project
MUBFS, PO Box 967, Fort Portal, Uganda

Andrew Plumptre, Albertine Rift Programme
Wildlife Conservation Society, PO Box 7487, Kampala, Uganda

Ian Redmond, GRASP Partnership
PO Box 308, Bristol BS99 3WH, UK

Vernon Reynolds , Budongo Forest Project
Orchard House, West Street, Alfriston, East Sussex BN26 5UX, UK

Elizabeth A. Ross, The Kasiisi Project
110 Concord Road, Weston, MA 02493, USA

Innocent B. Rwego, Kibale Eco-health Project
MUBFS, PO Box 409, Fort Portal, Uganda

Jane Southworth
Department of Geography, 3141 Turlington Hall, PO Box 117315, Gainesville, FL 32611-7315, USA

Jared Stabach, The Woods Hole Research Center
149 Woods Hole Road, Falmouth, MA 02540-1644, USA

Thomas Struhsaker
2953 Welcome Drive, Durham, NC 27705, USA

Rosie Trevelyan, Tropical Biology Association
Department of Zoology, Cambridge CB2 3EJ, UK

Dennis Twinomugisha, Kibale Fish and Monkey Project
MUBFS, PO Box 967, Fort Portal, Uganda

Melanie Virtue, Great Ape Survival Project (GRASP) Coordinator
GRASP Secretariat, UNEP, PO Box 30552, Nairobi, Kenya

Wayne Walker, The Woods Hole Research Center
149 Woods Hole Road, Falmouth, MA 02540-1644, USA

Elizabeth Williamson, Karisoke Gorilla Project
Department of Psychology, University of Stirling, Scotland, UK

Richard Wrangham, Kibale Chimpanzee Project
Department of Anthropology, Peabody Museum, Harvard University, 11 Divinity Avenue, Cambridge, MA 02138, USA

Klaus Zuberbühler , Budongo Forest Project
The School of Psychology, Westburn Lane, St. Andrews, Fife KY16 9JU, Scotland, UK

IAN REDMOND AND MELANIE VIRTUE

Foreword

In the closing years of the twentieth century, reports of ape populations in decline caused increasing alarm among conservationists. Not everyone was convinced at first, because broad trends were being extrapolated from patchy data. Many of the reports were anecdotal, and dealt with the fate of individual apes rather than populations; long-term research sites, however, yielded relatively accurate figures over time. Eventually, more and more eyewitness accounts from researchers, conservation field-workers, and investigative journalists drew the same conclusion: our closest relatives in the animal kingdom were facing extinction in a matter of decades unless the causes of their decline were addressed.

The causes were, and still are, human activities. Most of these – hunting, logging, agriculture, and warfare – have been practiced for millennia at self-evidently sustainable levels. The difference today is one of scale – especially when the activities are driven by international commerce and demand from the developed world for resources such as timber and minerals from ape habitats. Even natural threats such as disease are being exacerbated by the impact of the modern world on the apes' habitat. If these pressures continue unchecked, local extinctions will increase, leading to total extinction in the wild within our lifetime.

Attention was drawn in the 1990s to the rise of the commercial bushmeat trade in Africa, linked to the expansion of logging concessions into previously inaccessible forests, especially in the Congo basin (Redmond, 1989; Pearce and Ammann, 1995). Bushmeat – the meat of wild animals – varies from caterpillars to elephants. It provides the main source of protein for millions of people, and much of it is legal, though increasingly unsustainable in the face of modern hunting methods and rising demand in urban markets and restaurants. Great apes, however, are protected by law in every one of their 23 range states (countries in which they occur naturally). Unfortunately, wildlife law enforcement and prosecutions are rare in most range states, so this is little deterrent. In addition to direct hunting, apes are also maimed and killed by snares and traps set for other species, even in parts of Africa where ape-meat is not eaten. Moreover, there is also a less well-publicized trade in ape body parts (fingers, hair, genitalia, etc.) for use in traditional African medicine. Live infants are sometimes a lucrative byproduct of bushmeat hunting, but are also captured on demand. In short, apes are seen as a source of easy money in places with few job opportunities.

Similar fears were also being expressed for the orangutans in Borneo and Sumatra. There, the main threat was the loss and degradation of habitat from logging – much of it illegal – and conversion of forest to agriculture, coupled with the killing of crop-raiding adults and the capture of infants for the illegal wild animal trade (e.g., Galdikas, 1995; Rijksen and Meijaard, 1999).

In both Africa and southeast Asia, the very visible and photogenic problem of confiscated orphan apes drew the attention of the world's media, but less attention was given to addressing the underlying causes that brought these orphans into human care. Shocking images and a growing body of independent evidence galvanized many primatologists and non-governmental organizations (NGOs) into action. Among the many initiatives worldwide, the Ape Alliance was convened in 1996 as an international coalition of NGOs and individuals working for the welfare and conservation of all apes. Its first bushmeat report painted a bleak picture for African apes and made recommendations for governments, NGOs and timber companies (Ape Alliance, 1998). The Bushmeat Crisis Task Force was established in the USA and became a source of reliable information for policy makers (see, for example, BCTF, 2000). A prominent group of scientists surveyed all great ape research sites and revealed that 96% of them were recording declines (Marshall *et al.*, 2000). They successfully lobbied the US Government for support, resulting in the establishment of the Great Ape Conservation Fund. In Japan, primatologists raised awareness and funds through annual conferences and lectures organized by SAGA – Support for African and Asian Great Apes. But the decline seemed to worsen as new factors emerged – ebola outbreaks in Africa and forest fires in Borneo and Sumatra linked to illegal logging and forest clearance for oil palm plantations.

What was needed, it seemed, was a global strategy that unified and coordinated the many disparate efforts, a strategy that all concerned helped to develop and implement. After broad consultation amongst interested parties, the United Nations Environment Programme (UNEP) launched GRASP, the Great Ape Survival Project, in 2001, inviting each government to designate a Focal Point and begin preparing a national policy document to guide decisions relating to apes or their habitat. The following year at the World Summit on Sustainable Development, with the addition of UNESCO, GRASP became a "Type II Partnership," bringing together UN bodies, conventions, governments, NGOs, academics, and the private sector, as described by Ishwaran (Chapter 19).

Early efforts were hampered by lack of funds, but even at this stage, a significant proportion of funds raised was applied to field projects, to directly benefit the apes, their forests, and the human communities living in or around these forests. Links to sustainable development were critical because 16 of the 23 great ape countries are "Least Developed Countries" – defined by the UN as having a per capita income of under USD 800 per annum. Clearly, the message must get through that great apes (and their habitats) are great assets. In Uganda, for example, tourism is the third largest earner of foreign exchange, and 20% of ape-viewing permit fees are shared with surrounding communities. This "gorilla and chimpanzee money" has paid for new clinics, schools, community centers, and roads, and has created a positive attitude to protected areas hitherto coveted for agriculture. But tourism brings new risks, such as the potential for disease transmission between visitors and the apes, and easier poaching of habituated individuals. Moreover, tourism revenues can shrivel in the event of war or political unrest, and more than half of great ape range states have seen such upheavals in recent decades.

Nevertheless, as a result of GRASP efforts, there is now a high-level commitment to ensure great apes do survive. This is encapsulated in the Kinshasa Declaration and Global Strategy (UNEP, 2005), signed by 30 governments and all the relevant UN bodies, NGO partners, and a few private sector interests. Each partner plays to its own strength to help implement the Strategy. For example, the Convention on International Trade in Endangered Species of Fauna and Flora (CITES) has established a task force on great apes to help stem the international trade in infant apes, while the Convention on Migratory Species (CMS) is currently supporting the range states to negotiate a legally binding agreement on gorillas. In order to implement the Global Strategy, a Program of Action and an Activity and Finance Plan have been developed to attract new sources of conservation finance for apes and ape-friendly development projects. The latter aims to improve the lives of people who might otherwise have little option but to hunt apes or destroy habitat. In this way, GRASP aims to help achieve the Millennium Development Goals as well as secure a future for great apes.

Despite all the above efforts, however, the fact remains that the economic pressures destroying ape habitat are several orders of magnitude greater than current available conservation finance. There are some successes on record, most often associated with long-term research sites; not only does the study population benefit from a higher level of incoming resources, the world also hears about their plight through articles, books, and documentaries. The heroic efforts of outstanding individuals, however, can do little more than hold back the tide here and there. What is needed is an equivalent economic incentive to counter the clearing of land.

Ironically, this economic incentive may be an unexpected result of the new threat to the survival of all apes, human and non-human. Apart from rising sea-levels, human-induced global climate change seems likely to affect rainfall patterns. If rainfall patterns alter with rising temperatures, the ecological conditions that give rise to forests will change, causing forests to move or disappear from areas currently considered to be priority sites for great apes. This is most significant in areas of fragmented forest, where fences and other human obstacles will inhibit apes from crossing open areas. Reforestation of corridors between forests could ameliorate this risk, and also help to absorb atmospheric carbon. Protection of standing forests, or "avoided deforestation" should also be valued in a system of total carbon accounting, recognizing the carbon in soils and peat as well as wood. Many feel that the best way to achieve this is through private sector trading of forest carbon credits, bringing significant new resources for the sustainable management of forests (Swingland, 2003). GRASP is exploring ways to use carbon financing to fund the better management of forests in the 23 great ape range states, in particular, the 94 priority sites identified by the GRASP Interim Scientific Commission (GRASP Interim Scientific Commission, 2005) as the locations that contain the most important populations of all great ape taxa. Protecting these high conservation value forests will not only ensure a future for gorillas, chimpanzees, bonobos, and orangutans, it will bring employment and sustainable development to forest-dependent human communities, and, by keeping carbon in trees and underlying soils and peat, help to mitigate against climate change.

This new pro-forest economic pressure should win the support of those who wield power over land-use planning decisions, large-scale development projects, and wildlife law enforcement. Only if these key decision makers and the communities they serve understand the importance of apes and ape habitat will they take the necessary steps to ensure they survive. The role of long-term research projects, as this volume amply illustrates, is critical to this process, for without accurate information and scientific understanding how can informed decisions be made?

Great apes are not, however, just interesting research subjects. They are the gardeners of the forest – keystone species in the ecology of African and Southeast Asian forests, dispersing seeds, creating light gaps, and pruning branch-tips while feeding. In other words, the forests need the apes as much as the apes need the forests. And their habitat happens to comprise two of the planet's three major tropical forest blocks that are essential for global climate regulation and the global carbon cycle. More than a billion of the world's poorest people depend directly on these same forests and every person on the planet benefits indirectly. If we are serious about slowing climate change and achieving the Millennium Development Goals, we *need* those forests to continue providing ecological services for the whole of mankind. Ergo, we *must* find the resources to implement the Global Strategy – including scientific research and monitoring programmes – thereby ensuring healthy populations of apes, to maintain in perpetuity the health of the forests for the benefit of all.

REFERENCES

Ape Alliance (1998). *The African Bushmeat Trade: Recipe for Extinction.* London: Ape Alliance.

BCTF (2000). *BCTF Fact Sheet: The Role of the Logging Industry.* Washington, DC: Bushmeat Crisis Task Force.

Galdikas, B. M. (1995). *Reflections of Eden.* Boston, MA: Little, Brown.

GRASP Interim Scientific Commission (2005). Report available at http://www.whrc.org/africa/prioritypops/maps_data.htm.

Marshall, A. J., Jones, J. H., and Wrangham, R. W. (2000). The plight of the apes: a global survey of great ape populations. A briefing prepared for Representative George Miller and Representative Jim Saxton *Re: H.R. 4320.*

Pearce, J. and Ammann, K. (1995). *Slaughter of the Apes: How the Tropical Timber Industry is Devouring Africa's Great Apes.* London: World Society for the Protection of Animals.

Redmond, I. (1989). *Trade in gorillas and other primates in the People's Republic of Congo: an investigation for International Primate Protection League.* Summerville, SC: International Primate Protection League.

Rijksen, H.D. and Meijaard, E. (1999). *Our Vanishing Relative: The Status of Wild Orangutans at the Close of the Twentieth Century.* Dordrecht, Netherlands: Kluwer Academic Publishers.

Swingland, I. R. (ed). (2003). *Capturing Carbon and Conservation of Biodiversity: the Market Approach.* London: Earthscan Publications.

UNEP (2005). Report available at http://www.unep.org/grasp/Publications/Official_Documents/official_docs.asp.

Preface

Global climate change has always been a challenge to species survival and, of course, evolution and extinction are driven in large measure by adaptive strategies that are triggered by habitat pressure and the loss or gain of opportunities. Today, those of us concerned with nature conservation are confronted by the onset of massive and rather rapid environmental changes, accelerated if not caused by human activity, past and current. I do not think there has been any time in the past where, in the human context, climate change has had such enormous and recognized implications for a world that we arrogantly consider ours.

Those of us who concern ourselves with the broad task of species survival have relatively few tools at our disposal in the face of such a gigantic challenge. There are, however, some things that are being done and perhaps more critically, could be done. Sharing knowledge and collaborating on an interdisciplinary and international basis is fundamental. Science may not have all the answers but it certainly offers a framework for strategic planning.

In some cases, perhaps little can be done to stop habitat loss but I sincerely believe that the concept of conservation remains valid and much that is going wrong can, in fact, be put right. Forests are particularly important and, along with wetlands, are probably the most important for the consideration of policy makers and the population at large. The misuse and destruction of forests for commercial purposes is a major threat and the need for energy, be it biofuels or charcoal is a real issue that has to be tackled now rather than tomorrow. Poverty is cited as a key factor and, whilst this is true, forest destruction will aggravate and increase poverty in the long term. Only when we know the facts can we contemplate the way ahead.

In this publication, a number of experts have been brought together to offer knowledge and experience gleaned from work across a broad range of forest habitats in Africa. The forests of southeast Asia and the New World are different, of course, but the African experience will surely be relevant and of interest to conservationists in these other regions.

This volume, which celebrates the twentieth anniversary of the Kibale Chimpanzee Project in Uganda, is an important and timely contribution and it will be of great benefit to all of us who are so engaged with the challenges of nature conservation in the twenty-first century. The editors and authors have to be congratulated, not only for the volume but also for their work in the field.

Richard Leakey, FRS

Acknowledgments

For financial support of the Kibale Chimpanzee Project we thank the following funding agencies: The National Science Foundation, The National Geographic Society, The Leakey Foundation, The Getty Foundation and Harvard University.

We thank the following for their financial support of the Kasiisi Project: the James and Gloria Stewart Foundation, the Seymour and Julia Gross Foundation, the people of the Town of Weston, MA, especially the staff, children, and parents of the Weston Public Schools who have enthusiastically supported the Kasiisi Project for 10 years, and the congregation of First Parish Church who have generously donated to the project since 1999, and who have loyally bought more Ugandan baskets than anyone could possibly use in a lifetime. Thanks go also to the Friends of the Kibale Chimpanzee Project and The Kasiisi Porridge Project, UK for supporting the work of the Kasiisi Project. For advice and fundraising assistance we thank Cultural Survival and Wildlife Concern International.

Most of all we would like to express our gratitude to the teachers, parents and children of Kanyawara, Kasiisi, Kigerama, Kiiko, and Rweteera primary schools who immeasurably enriched our lives by allowing us to be, for a short time, a part of theirs.

We thank the following, who by providing research permission and support, made our time in Uganda possible: the Uganda Wildlife Authority, Makerere University Biological Field Station, and the National Council for Science and Technology.

For financial support of the Kibale Chimpanzee Project 20th anniversary workshop on long-term research and conservation we thank: The Alexander von Humboldt Foundation, The German Federal Ministry of Education and Science, and Brian Hare. We are grateful to all those who took time out of busy lives to join us in celebrating our twentieth anniversary in Uganda and whose contributions and participation were the basis of this book. We thank Kayo Burmon for designing the book cover.

We feel enormous gratitude for the kindness and flexibility of the people of Kanyawara who by welcoming our family to their country and forgiving our clumsy attempts to fit in made Uganda, for 20 years, a second home. We remember especially those who died during this time and who contributed so much to the success of our work: graduate students, field assistants, domestic staff, and MUBFS employees.

We owe special thanks to Tom Struhsaker, who first suggested that Kibale Forest would be a good place to combine chimpanzee research and small children, and to Gil Basuta and John Kasenene who made our time at MUBFS happy and productive.

And last, but not least, we would like to acknowledge the individual chimpanzees of the Kanyawara community, who brought us to Uganda in the first place. We hope that this book will be a contribution to their survival.

RICHARD WRANGHAM

1

Why the link between long-term research and conservation is a case worth making

In 1871 Charles Darwin feared for the future of great apes. "At some future period, not very distant as measured by centuries," he wrote, "the anthropomorphous apes ... will no doubt be exterminated" (Darwin, 1871, p. 891). While Darwin's prognostication might seem gloomy, to those concerned with the conservation of great apes Darwin seems optimistic to have anticipated extinction in centuries rather than decades. The contemporary threats to tropical forests are so numerous and intense that most conservationists would be delighted if they could be assured that orangutans (*Pongo pygmaeus*), chimpanzees (*Pan troglodytes*), bonobos (*Pan paniscus*), and gorillas (*Gorilla gorilla*) would all still survive in the wild in 2100. Unfortunately, however, even such a modest hope may be unrealistic. The three most threatened species, orangutans, bonobos, and gorillas, are widely considered as candidates for global extinction within the next 100 years (Beck *et al.*, 2001; Miles, 2005). If they go, so will large numbers of other animals and plants.

The problem would be bad enough if the scale of the threats to which tropical forests are currently exposed were to continue unchanged in the near future. All indications are, however, that the challenges of maintaining forests are going to grow enormously. This means that, if the tidal wave of forest destruction is ever to be turned back, a critical question is how much will be lost before then. What we do now will substantially affect the answer.

Under these circumstances it is the responsibility of conservationists to explore every avenue for saving the forests. In this book we highlight an approach that we believe has had less exposure than it deserves. In conversation, tropical field scientists often suggest that the establishment of a long-term research program helps to protect the gazetted area in which it is located. Occasionally, the rationale for such opinions has been made explicit (Plumptre and Williamson, 2001; Pusey *et al.*, 2007), and research stations have sometimes been advocated as a way to help preserve specific populations (e.g., Morgan and Abwe, 2006). But the potential conservation benefits of long-term field stations do not yet appear to have been widely discussed in a public forum. Nor have Protected Area Authorities (PAAs) or conservation organizations consistently worked to recruit or encourage the establishment of research stations on the premise that they might aid the goals of the PAA. While many reasons make such omissions understandable, the result is that the potential for taking advantage of positive conservation outcomes from long-term research may have been under-used.

AIMS

The aim of this book, therefore, is to illustrate in detail some of the ways in which field stations have contributed to conservation, with the ultimate goal being to encourage the establishment and maintenance of such stations as part of an enlightened conservation strategy. The book has two main sections, the first of which explores the relationship between research and conservation in a particular location, Kibale National Park, Uganda. As several chapters outline, in 1970 Thomas Struhsaker began studying the ecology and behavior of red colobus monkeys in Kanyawara on the northwest edge of what was then the Kibale Forest Reserve. Over time he attracted others to the same site, and in 1989 a project that had begun with a single small hut became the Makerere University Biological Field Station, which was to have multiple buildings including researcher housing, laboratories, a lecture room, and a capacity to house 50 or more visitors. Since then, there has been a growing series of long-term projects concerned with primates, fish, butterflies, trees, and other taxa. Many of the researchers brought in by these studies have participated in projects that are directly or indirectly relevant to conservation. Chapters 2 to 12 present the relationship between research and conservation in a Ugandan context by surveying the diversity of research projects and evaluating their significance. Authors of several chapters also discuss how conservation efforts might be improved in the future. This section thus provides a case study describing how numerous conservation consequences, from forest restoration to ecotourism and community development, arise from activities that began entirely as pure research. A typical example of the intersection of research and conservation is demonstrated by the Kibale Snare Removal Project, a result of the research of the Kibale Chimpanzee Project (see Box 1.1).

Box 1.1. Kibale Snare Removal Project

Amy Pokempner

Throughout Uganda, over 25% of individuals in habituated communities suffer snare related injuries (Waller *et al.*, 2001; Wallis *et al.*, 2002; Plumptre *et al.*, 2003). Although primates are not actively targeted for bushmeat in Uganda, chimpanzees may often fall victim to snares intended to capture small mammals such as duiker and bushbuck. Snares may range from hand-fashioned metal and nylon spring traps to more powerful steel leg traps that are difficult to detect in the forest, particularly for a large social animal such as a chimpanzee. In the Kanyawara community alone, 35% of individuals show the lasting signs of wounds in the form of crippled or missing hands, feet, and digits. In at least three of these cases, these injuries have resulted in an early death. A study in 2000 by KCP estimated a density of 15,000 snares set in Kibale National Park at any given time resulting in a 3.7% risk of a chimpanzee being snared each year (Wrangham and Mugume, 2000).

In response to this threat, the KCP developed the Kibale Snare Removal Project (KSRP) based on the expertise and dedication of two former park rangers, John Okwilo and Aloysius Makuru. Within the first years of the program (1997 to 2000) the KSRP removed 2290 snares, averaging a collection of 67 snares per month. While patrols initially focused on the area surrounding Kanyawara, they soon expanded to cover areas throughout the park, including Ngogo and Kanyanchu, as well as the particularly vulnerable regions along the park boundary. Within recent years, average collection has dropped to 28 snares per month as patrols have continued to increase. Since the initiation of the project, UWA has also intensified its ranger-based monitoring efforts, working in close coordination with the KSRP. With a wealth of long-term experience, the KSRP staff provide essential training not only to UWA rangers but also to rangers in Budongo and Kalinzu where similar snare removal projects have since been set up. KSRP rangers also collect data on all illegal activities as well as on any animal signs they encounter, thus providing data that can be incorporated into UWA's ongoing monitoring system (MIST: Management Information System) and that help identify trends within the park for more effective management.

In Chapters 13 to 18 the second main section of the book presents evidence that the Kibale experience, of a "pure" research project leading to multiple conservation benefits through a variety of locally appropriate initiatives, happens commonly. Six major sites for studying chimpanzees and gorillas illustrate the point by showing that, in every case, what began as a behavioral study has become a multi-dimensional effort to persuade the local people and governments that conservation is worthwhile. The consistency of such effects might seem remarkable, given that the spin-off projects described in this book have often been developed by people who had no relevant background in conservation or community relations. But the surprise is easily explained. However rarefied a biologist's intellectual goals may be, few researchers can spend long in a tropical forest without becoming drawn to its inevitable conservation problems. Perhaps researchers hear the chopping of trees, or find snares, or watch trucks of bushmeat leaving the forest. They might fear for the future of their studies, or for particular threatened species, or maybe for the global loss of biodiversity. Some see the unique position in which their scientific efforts have placed them as creating a responsibility. Others find personal satisfaction in engaging with local people to help stop them from ruining their local and national environmental services. Just as important as the researchers are their families, who likewise are repeatedly drawn to help solve problems related to conservation and community. Whichever of these or many other factors serve to motivate the inhabitants of a field station, history tells us that very often, long-term visitors get involved.

Unfortunately, although the conservation benefits that emerge from research stations appear to be important, there is no decisive evidence on the question, because the benefits have been examined so little that it is still not possible to convincingly distinguish cause and effect from other kinds of correlation.

For example, following the establishment of a research station, the area containing it has often been awarded a higher level of protection. Thus Jane Goodall began work in Gombe Stream Reserve in 1960; her writings drew national and international attention to the site; and the area was declared a National Park in 1968. Similarly, Kibale was a Forest Reserve until 1994, when it became a National Park following years of advocacy for its conservation. Sometimes, the higher level of protection follows directly from intense efforts made by researchers. Ten years after beginning research in the Mahale Mountains in 1965, Toshisada Nishida and Junichiro Itani presented a substantial proposal to the Tanzanian Government to protect the area. They were encouraged to produce a detailed conservation plan, and this led to the gazettement of Mahale as Tanzania's eleventh National Park in 1985.

In such cases it seems likely that the researchers' presence, and in some cases their direct efforts, have helped in generating governmental interest and hence their legislative support. But skeptics might note that the sites in which scientists tend to set up research stations are also likely to be the kind of habitat most worthy of a high level of protection. In other words, Gombe, Kibale, and Mahale could, in theory, have achieved their high level of protection whether or not there had been a research station. It will need a careful research design that compares forests differing by the presence or absence of a field station to resolve this question satisfactorily. Such a study has not yet been done. But pending such a study, we believe there is already sufficient informal evidence to encourage the promotion of long-term field stations as a practical research strategy. Hence this book.

The rationale for this book thus has two main components. The threats will grow enormously, and long-term research stations tend to have positive effects. I briefly elaborate on these points.

THREATS TO THE TROPICAL FORESTS

Since the middle of the twentieth century, an increasing proportion of natural areas throughout the tropics has been protected, giving hope that a sizeable number of habitats and species will survive in the near future. But however strong the legislation that currently supports Protected Areas (PAs), their effectiveness for conservation can never be guaranteed.

Most Protected Areas were largely uninhabited at the time when they were gazetted. It was, of course, wise of conservationists to negotiate for as much land as possible before it was occupied. But the political forces that challenge the wisdom of preserving land for its conservation value have until now been weak in most tropical countries. We can expect radical change in the future, based on the expectation that the intensity of competition over land use will rise sharply.

Two pressures are particularly alarming. First, population growth in tropical countries is generally high, and in some cases such as Uganda it is startling (Goldman *et al.*, Chapter 12). All environmental services are placed in threat as populations grow. The scale of the growth is high. In Uganda it has not yet reached the point at which people find it hard to obtain land, but it may not take long before local politicians complain about the protection given to Forest Reserves or National Parks.

Second, rising economic expectations are creating acquisitive trends in second-world as well as in third-world countries. Clearly, the developed world cannot reasonably complain about the desire of China, India, Brazil, and other large and rapidly growing economies to aspire to a standard of living enjoyed to date by Europe, Japan, and North America. But, regardless of the moral justification for growing economic achievements, the immediate prospects for the natural resources of range countries look devastating. Already, China's demand is fueling timber extraction in West Africa. Tropical habitat countries will doubtless be offered significant political or economic rewards by more powerful nations competing for scarce resources. The temptations offered are easily illustrated.

Take Uganda. President Yoweri Museveni came to power in 1986 and almost immediately expressed strong support for conservation in forests. In the next 20 years he and his government consistently supported habitat protection, including the establishment of a series of National Parks in the early 1990s (Bwindi Impenetrable National Park, Mgahinga National Park, Kibale National Park, Ruwenzori National Park, Lake Mburo National Park). Museveni's enlightened policy apparently bore fruit. By 2004, tourism had become the principal internal source of foreign exchange, and ecotourism to see gorillas and chimpanzees was responsible for 52% of the tourist revenue. Museveni affirmed his support for forests in a public forum in Uganda at the 2006 International Primatological Congress. "Lead us not into temptation," he said, "deliver us from evil."

Yet within a year Museveni was fighting a battle on behalf of an Indian industry to sacrifice almost a quarter of Mabira Forest, one of Uganda's most important Forest Reserves, for the sake of growing sugarcane. He did so despite widespread national opposition including a public protest on the streets of the capital, Kampala, which turned violent in April 2007 and led to the loss of three lives. Museveni argued that, by turning over 71 sq km of forest for industrial sugar-cane production, he would provide 3500 jobs and earn the country around US$7 million. Responding to his critics, he said he would "not be deterred by people who don't see where the future of Africa lies." At the time of going to press, the fate of Mabira remains uncertain.

Mabira provides a warning. Even the most enlightened presidents and governments will come under increasing pressures to utilize natural lands for short-term gain. We have been living in a state of relative ease. The pressures on Protected Areas in the tropics have barely begun, especially for forests that offer little direct financial reward in terms of tourism.

THE POSITIVE EFFECTS OF FIELD STATIONS

Conservation biology is a rapidly growing and important academic field, and although it is concerned primarily with scientific understanding it also pays attention to the practical aspects of protecting habitats and assessing success. Yet even so, there has been little published consideration of such "fuzzy" factors as the value of personal relationships between conservationists and managers, or the importance of long-term commitment, in promoting habitat management. As several chapters in this book note, more interaction between biologists and social scientists is desirable in the context of understanding the perspective of local communities. The same can be said for the question of how effectively field conservationists manage their "political" agendas. Kasenene and Ross (Chapter 10) note that, in practice, personal relationships matter, all the more so because cultural differences can easily interfere with the development of trust. These are critical issues if people from the developed world, who might be naïve, impatient and/or idealistic, are to interact effectively with managers in tropical countries.

The encouraging evidence is that conservation efforts can work even under difficult circumstances. A striking example is the case of mountain gorillas (*Gorilla gorilla berengei*). From a low point in the early 1980s, the number of mountain gorillas has risen steadily during the last two decades, thanks to cooperative efforts from Rwanda, Uganda, and the Democratic Republic of the Congo as well as a variety of organizations coordinated by the International Gorilla Conservation Program (Williamson and Fawcett, Chapter 18). During this period the local human populations have been subject to war and genocide, and the habitat has been under pressure from pyrethrum producers, farmers, and charcoal-burners. Among the factors helping the mountain gorillas, the most obvious is ecotourism, which was initiated out of the behavioral research program and led to strong support from the government for the gorillas. While the mountain gorillas in the Virunga Mountains and Bwindi Impenetrable National Park have fared well, during the same period gorillas in various sites in the Democratic Republic of the Congo have suffered badly (Stoinski *et al.*, 2008).

Unusually, the total population of mountain gorillas has been counted repeatedly, allowing their real conservation status to be assessed. More often conservation activities are carried out without such direct evidence of their consequences. Since there are dangers as well as benefits from the presence of long-term researchers, as Boesch *et al.* (Chapter 16) note, such evidence is desirable. But even without much formal data on conservation outcomes, the following chapters suggest that long-term research, however abstract in its initial conception, leads to beneficial results with remarkable consistency. Overall, they suggest that long-term research stations tend to evolve into invaluable foci of interaction among managers, biologists, social scientists, and conservationists. In a world threatened by massive habitat loss, therefore, a more sustained effort to promote the development of field stations may be a promising way to help Protected Area Authorities in their mission in general, and in specific locales to defer the time when "the anthropomorphous apes will be exterminated."

REFERENCES

Beck, B. B., Stoinski, T. S., Hutchins, M. *et al.* (2001). *Great Apes and Humans: The Ethics of Co-Existence*. Washington DC: Smithsonian Institute.

Darwin, C. 1871 (2006). *The Descent of Man, and Selection in Relation to Sex*. New York: W. W. Norton.

Miles, L. (2005). *World Atlas of Great Apes and Their Conservation*. Berkeley, CA: University of California Press.

Morgan, B. J. and Abwe, E. E. (2006). Chimpanzees use stone hammers in Cameroon. *Current Biology*, **16**, R632–R633.

Plumptre, A. and Williamson, E. A. (2001). Conservation-oriented research in the Virunga Region. In *Mountain Gorillas: Three Decades of Research at Karisoke*, ed. M. M. Robbins, P. Sicotte, and K. J. Stewart. Cambridge: Cambridge University Press, pp. 361–389.

Plumptre, A. J., Arnold, M., and Nkuutu, D. (2003). Conservation action plan for Uganda's chimpanzees. Workshop report, Wildlife Conservation Society and Jane Goodall Institute, 2003–2008.

Pusey, A. E., Pintea, L., Wilson, M. L., Kamenya, S., and Goodall, J. (2007). The contribution of long-term research at Gombe National Park to chimpanzee conservation. *Conservation Biology*, **21**, 623–634.

Stoinski, T. S., Steklis, H. D., and Mehlman, P. T. (2008). *Conservation in the 21st Century: Gorillas as a Case Study*. New York: Springer.

Waller, J. C. and Reynolds, V. (2001). Limb injuries resulting from snares and traps in chimpanzees (*Pan troglodytes schweinfurthii*) of the Budongo Forest, Uganda. *Primates*, **42**, 135–139.

Wallis, J. J., Munn, J., and Reynolds, V. (2002). Snare injuries in chimpanzees: collateral damage of the bushmeat trade. *American Journal of Primatology: Program and Abstracts of the Twenty-Fifth Annual Meeting of the American Society of Primatologists*, 1–4 June 2002, 57.

Wrangham, R. W. and Mugume, S. (2000). Snare Removal program in Kibale National Park: a preliminary report. *Pan Africa News*, **7**, 18–20.

MOSES MAPESA

2 Links between research and Protected Area management in Uganda

INTRODUCTION

As executive director of the Uganda Wildlife Authority (UWA), I see two principal ways in which biological research aids conservation. First, are various processes that originate independently of UWA's direct involvement, such as the generation of data and proposals from university field stations, many of which are discussed elsewhere in this book. Second, is the production of data specifically aimed at assisting UWA's management aims. This chapter deals with the latter.

UWA was established in 1996, as a semi-autonomous parastatal governed by a Board of Trustees. It is in charge of the management of ten National Parks, twelve Wildlife Reserves and seven Wildlife Sanctuaries, and it provides guidance for five Community Wildlife Areas. Each of these Protected Areas (PAs) is allocated to one of seven administrative areas: the Kibale, Queen Elizabeth, Bwindi-Mgahinga, Murchison Falls, Mt. Elgon, Kidepo, and L. Mburo Conservation Areas (Fig. 2.1).

UWA operates under a Strategic Plan which guides management interventions. This strategic plan is roughly divided into three strategic programs: Protected Area management, community conservation and benefits, and wildlife management outside Protected Areas. The monitoring and research program is part of the PA management program.

In this chapter I highlight the monitoring and research strategic program to illustrate its importance for conservation goals. Reasonably accurate data are vital for planning and implementing practical management policies, so research plays a critical role.

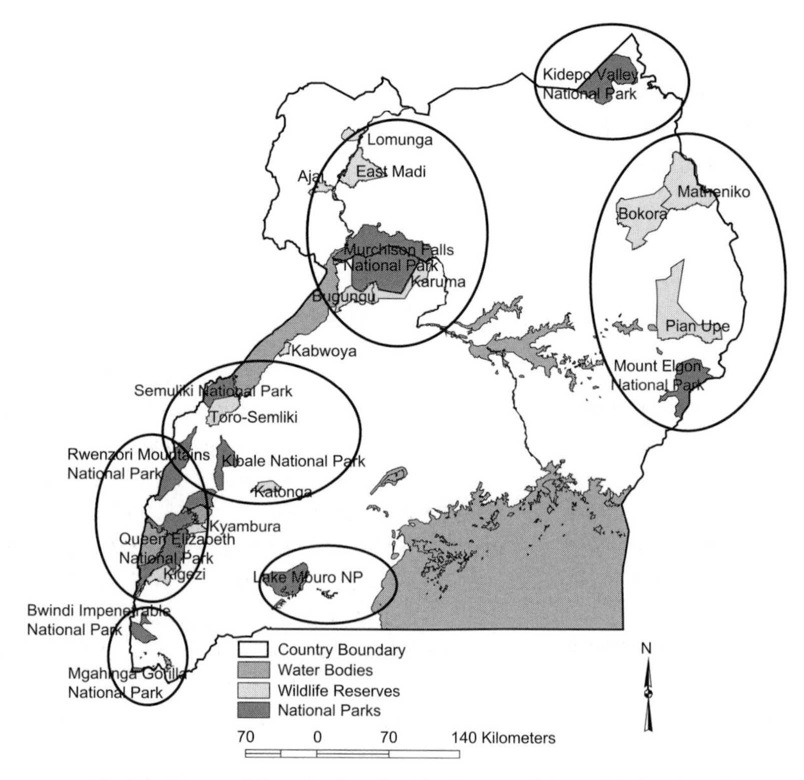

Fig. 2.1. A map of Uganda showing the Protected Areas (National Parks and Wildlife Reserves) managed by the Uganda Wildlife Authority. The seven administrative areas are circled.

RESEARCH AND MANAGEMENT IN THE PROTECTED AREAS

If we trace the link, in Uganda, between research and the management of Protected Areas, we find that, even prior to the creation of the National Parks, data were being collected on populations and distribution of wildlife. This research covered not only the relationship between wildlife and the environment, but also the interaction between wildlife and people. Based on this research, decisions were made about land use practices and many of the National Parks and Conservation Areas were created as a result.

By the 1950s, several Cambridge University researchers were working in the then Queen Elizabeth Conservation Area, studying habitat changes and animal carrying capacity. Intensive studies in the Queen Elizabeth Conservation Area and adjoining regions led in 1959 to the establishment of the first Institute of Ecology in Africa. The research findings from this institute were used for management actions during the 1960s and 1970s. An example was the management decision to cull hippos and elephants in Queen Elizabeth and Murchison Falls National Parks. Research data had shown that the large population of these animals had led to damage to the environment and needed to be reduced.

Renamed the Uganda Institute of Ecology (UIE) in 1971, this research station became the research arm of the Uganda National Parks. At that time, it largely focused on the savanna parks, including animal and plant ecology, tourist development, and the interaction between the National Parks and the local people. UIE played a lead role in the research into the animals and plants of Uganda and among other things conducted regional censuses and aerial surveys. Later, in the 1980s and 1990s, other research stations were established. Makerere University Biological Field Station in Kibale Forest and the Institute of Tropical Forest Conservation in Bwindi are both forest field stations, which generate data on forest-related ecosystems. Central to this research was the study of a number of key wildlife species, particularly primates. These included Uganda's great apes, chimpanzees and gorillas.

While initially much of the data generated was focused on basic and academic issues, information was also generated for management purposes. Results from some long-term studies, e.g., those of chimpanzees and gorillas, resulted in active tourism programs, which have become internationally popular with the attendant conservation benefits for these great apes. The management strategies for Kibale, Bwindi Impenetrable, and Mgahinga National Parks are informed by this data collection. The rules governing the behavior of tourists visiting these greatly endangered species are guided by the results of continuing research; examples include rules about the disposal of refuse and the distance that must be maintained between people and apes to protect the health of the animals. This is a very clear demonstration of the impact that long-term research programs have had on the management of the Protected Areas and on the conservation of their animals.

UWA now has a comprehensive Monitoring and Research program, which is centrally housed at the UWA headquarters in Kampala. The program is governed by the wildlife monitoring and research policy. Its clearly defined goal is to promote the collection and use of accurate, and timely information relevant to the conservation and management of Uganda's wildlife resources. Priorities for the Protected Areas are set and carried out, based on management strategies and criteria that are reviewed regularly.

The UWA's research priorities reflect the organization's mandate of conservation and development under four main themes: ecology, biodiversity, tourism, and socioeconomics. Similarly, monitoring activities are prioritized under the fields of ecology, socioeconomics, and development.

An in-house management system (MIST) has been developed to ensure the flow and dissemination of data from research and monitoring programs. This information is collected through ranger-based data collection (RBDC) and community-based data collection (CBDC), and from independent research. Every time rangers go out on patrol, they collect data on the distribution and number of key wildlife species and the distribution and frequency of illegal activities. These somewhat opportunistic data have three main aims. They complement long-term research programs; they help to guide the direction of long-term research; and they are used to compile the reports that help management interventions and operational planning. For example, a mammal report produced by the rangers while out on patrol shows not only the various sightings of different animal species but is also a measure of the location and area of the park that has been covered during a given patrol. This helps with resource allocation when undertaking Annual Operations Planning.

HOW HAVE WE USED RESEARCH INFORMATION WITHIN OUR PLANNING EFFORTS?

All the National Park's General Management Plans have been generated based on results from long-term research programs, supplemented by UWA ranger- and community-based data collection (Table 2.1). Research and monitoring data have helped in wildlife translocations, for instance when giraffes *(Giraffa camelopardalis rothschildii)* and elands *(Tragelaphus oryx)* were translocated to Kidepo. The rhino *(Diceros bicornis* – Black rhino, *Ceratotherium simum* – White rhino), extinct in Uganda for 20 years, has been reintroduced. Data from a number of studies were used to facilitate this initiative.

Monitoring and evaluation of tourist programs uses the results of long-term research. In Kibale National Park there is an ongoing study on the impact of tourism and habituation on the behavior of chimpanzees. Similar studies have been undertaken in Bwindi Impenetrable National Park on the gorillas. This kind of information is particularly important in planning for the great apes, given their close relationship to people. Information on group behavior has provided the basis for rules on the number of people in a tourist group and how close they can come to the

Table 2.1. *Ways in which research information is used by the Uganda Wildlife Authority*

Management plan	Research data
Preparation of General Management Plans for protected areas	Wildlife distribution, vegetation and socioeconomics
Tourism development and diversification	Habituation, health, evaluation of tourist impact
Management of wildlife/human conflict	Crop raiding, dangerous animals
Management of invasive species	Mapping and inventory
Development of specific action plans e.g., National Great Ape Survival Plan	Great ape distribution and density
Community agreements and concessions	Rate of resource exploitation
Implementation of Wildlife Use Rights Program e.g., setting hunting quotas	Wildlife population density and reproduction

animals. Continuing research on the impact of tourism on the apes helps UWA to modify these regulations at appropriate times.

Research information is used to address a variety of management challenges. For instance, crop raiding by wild animals is a perennial problem. Using research data, UWA has been able to give advice on the best crops to be grown in agricultural areas bordering the National Parks. In addition, experiments have tested a number of mechanisms which have the potential to reduce crop damage by keeping animals inside the Protected Areas, e.g., living thorny fences and trenches. Another problem for UWA that is assisted by research data is that of the control and management of invasive plant species. UWA have also developed specific action plans for individual animal species based on their conservation status. For example, the National Great Ape Survival Project has been designed to protect chimpanzees and gorillas.

Research information is used to plan ways that the communities living around National Parks can access some of the resources within the park. Very high population densities in some of these places means that there is often conflict between the people, who wish to access some of the resources in the parks, and the Protected Area authorities which protect them. In order to minimize illegal activities such as poaching, programs have been developed with the communities, which allow them access to some resources within the parks. This is successful so long as the level of resources harvested does not compromise the conservation objectives of UWA. In Bwindi Impenetrable National Park, research results have helped determine sustainable harvesting and this has improved the management of multiple use areas. Again, continuous long-term data are critical in ensuring that this program works well.

Concessions have been granted for management of some PAs in collaboration with the private sector and local government. Information and data are needed to negotiate these concessions and to agree on appropriate management intervention. A Wildlife Use Right program is being implemented, where some trade in wildlife species is permitted. This program is especially designed to manage animals living outside Protected Areas, which are threatened because of land use practices. By allowing this trade, UWA ensures that communities appreciate that wildlife is valuable and should be protected. An example is the sport hunting concession which UWA is piloting around Lake Mburo Conservation Area. Continuous monitoring and research are needed here to ensure that appropriate quotas are set and that the hunting does not lead to extinction.

SUMMARY

In conclusion therefore, UWA realizes the importance of long-term research in the management of Uganda's Protected Areas. Data collected are used extensively for decision making and for planning. UWA will therefore continue to ensure that there is a climate in Uganda that encourages long-term, short-term, and opportunistic research activities. We will support research, not only within the Protected Areas, but also within the adjoining areas that have an impact on the management of these Protected Areas. We will continue to undertake active monitoring and research programs in-house but also in collaboration with research institutions and field stations. We also recognize that one of the benefits to come from long-term research programs is improved capacity building within UWA, resulting from the number of UWA senior staff who have had experience working on long-term research programs. Moving into management positions within the organization, they bring with them an experience of the advantages of research that enhances their ability to work with research institutions and to get the most out of the data they receive.

WILLIAM OLUPOT AND ANDREW J. PLUMPTRE

3

The use of research: how science in Uganda's National Parks has been applied

INTRODUCTION

Uganda has a long history of ecological research. Some of the world's oldest permanent sample plots occur in its forests, dating back to 1933 (Eggeling, 1947; Sheil, 1996). Management of forests for timber was always aimed at making logging sustainable, and in this respect Uganda was well ahead of its time. Initially, research in forests was designed to improve forest management, leading to changes in management practices that even influenced tropical forest management in other countries around the world (Dawkins and Philip, 1998). Similarly, research on savanna ecology and wildlife dates back to the 1950s in Uganda, although tracking of wildlife populations and culling of elephants (*Loxodonta africana*) began as early as the 1920s. The establishment of the Nuffield Unit of Tropical Animal Ecology (NUTAE) in Queen Elizabeth National Park in 1961, which later became the Uganda Institute of Ecology (UIE), was well ahead of most research stations in other savanna parks in Africa. Some of the first studies of large mammal ecology in African savannas and the impacts of grazing/browsing and fire were made in Uganda's parks and led to management recommendations for culling in the late 1960s.

Most of the research in Uganda's Protected Areas has been associated with research stations. Uganda currently has three research stations in forested ecosystems: Makerere University Biological Field Station (Kibale National Park), Budongo Forest Project (Budongo Forest Reserve), and the Institute of Tropical Forest Conservation (Bwindi Impenetrable National Park). The Uganda Institute of Ecology ceased to function in the late 1990s because of lack of funding but it has the potential to be revived.

The boom of research in Uganda's Protected Areas during the early years was followed by a period of decline when Idi Amin took over the country in 1971. The period between 1975 and 1980 was a particularly bad one, as research in many sites virtually ground to a halt. Large mammal populations crashed at this time as a result of heavy poaching, particularly elephants (Fig. 3.1), which were killed for ivory, and as a result Uganda was not as interesting or as safe to visit as a researcher. Research increased again during the late 1980s, but was mainly driven by individual researchers (both national and international) studying for Masters and Doctoral degrees. There was much less research initiated by scientists or institutions aiming to answer questions with longer-term data. Only a few researchers who maintained a long-term presence in a forest were able to address some of these questions. In the 1950s–1970s a fair amount of research was driven by the needs of the management institutions. NUTAE and UIE produced data for the savannas, and the forest department's research unit for the forests. From the early 1990s to the present, much of the research has been based on the interests of individual researchers. As such, it has not specifically targeted all of the research needs of the Uganda Wildlife Authority (UWA) or the National Forest Authority (NFA).

This chapter is based on a much larger review of published research from Uganda's ten national parks and the Budongo Forest Reserve; currently the country's largest Forest Reserve. The review references 209 publications for the savannas and 300 for the forests, a small fraction of the publications that were reviewed. It addressed the need to show the conservation perspectives of different kinds of research in order to make findings from the different research areas readily interpretable for application to management.

Nobody really knows what is needed for the successful long-term management and conservation of Protected Areas, although great strides have been made in the last 50 years. Protected Area management is a long-term experiment that needs to be supported by research. Both basic and applied research projects are essential for managing Protected Areas, although conservation programs use applied research more directly. Basic research deals with defining ecosystem components and understanding how they work without specific applications in mind. It provides the foundation for predicting the impacts of threats and management strategies addressed by applied research. Applied research takes findings from basic research further by producing results that may be applied to real world situations, for example, in determining and quantifying threats, prescribing actions, and assessing the efficacy of the actions taken to address the threats. Long-term basic research often leads to applied research as researchers gain insights about threats and ways to address

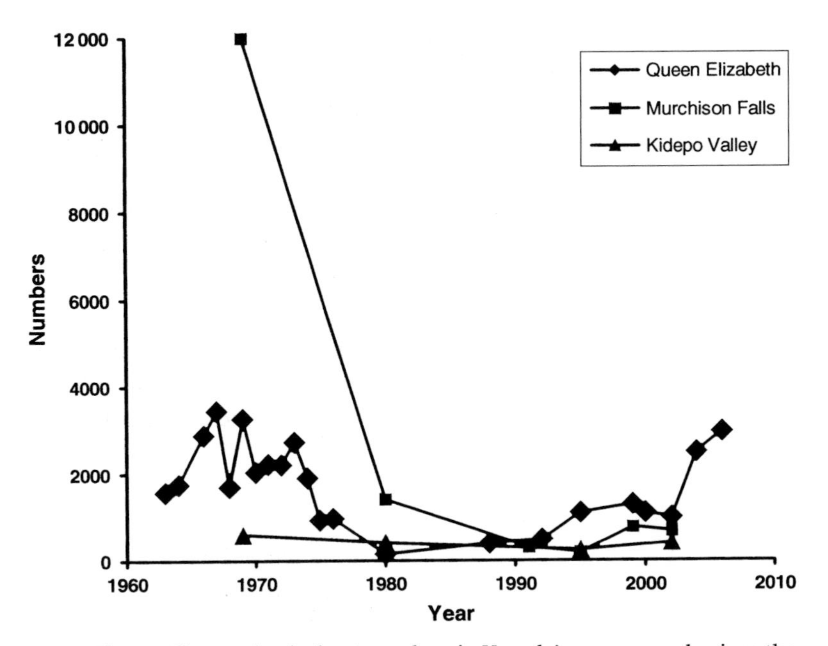

Fig. 3.1. Changes in elephant numbers in Uganda's savanna parks since the 1960s.

them. In this chapter we look at the long history of research in both the forests and savanna ecosystems in Uganda and assess how research has been used to change management interventions over this time.

RESEARCH IN SAVANNAS

Savanna research in Uganda has taken place in the four savanna parks: Queen Elizabeth, Murchison Falls, Lake Mburo, and Kidepo Valley National Parks (Fig. 3.2). Most of the research though has been concentrated in Queen Elizabeth because of the presence of NUTAE and then UIE. Research in the 1950s and 1960s focused on the ecology of large herbivores, particularly elephants and hippos (*Hippopotamus amphibious*) to better understand their population growth rates, reproduction, and ecology (Laws *et al.*, 1975; Eltringham, 1999) and their impact on the savannas. Together with these were studies of the impact of fire on the savanna ecosystem and the combined effects of the large mammals and fire. These studies showed that elephants and hippos had a major impact on vegetation when at high densities, and decreased plant diversity. In the case of hippos this was due to trampling, leading to erosion and areas of bare

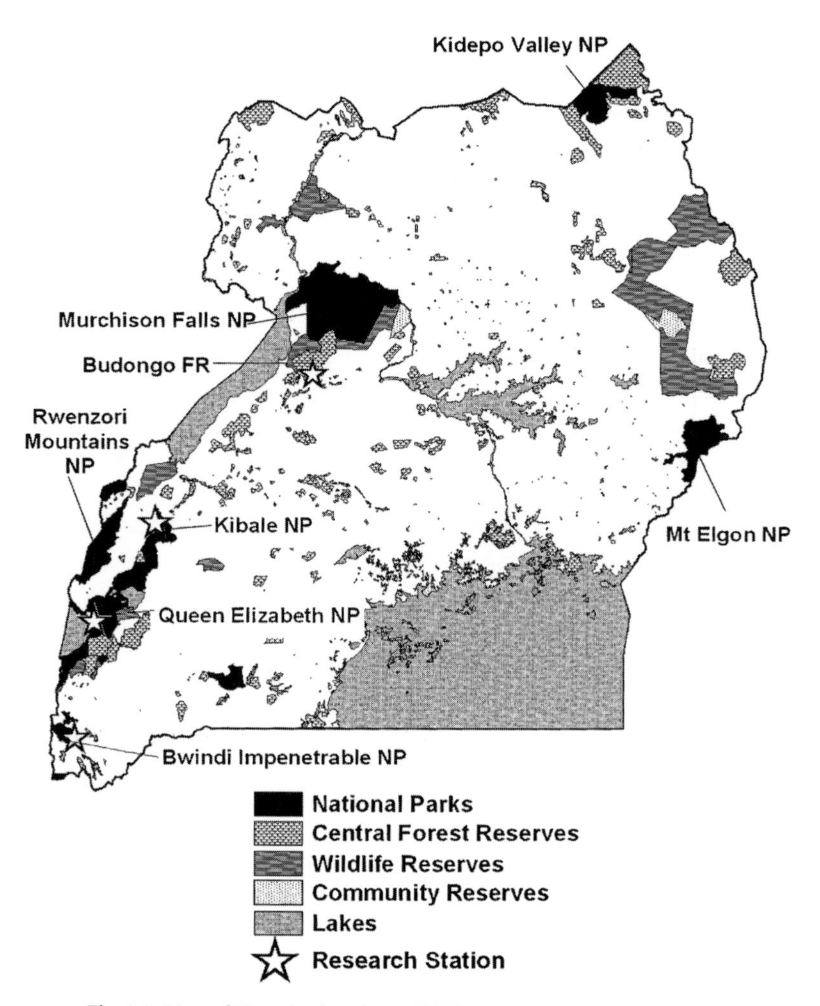

Fig. 3.2. Map of Uganda showing wildlife Protected Areas and research stations.

earth. Overgrazing by hippos also reduced fuel for fires, resulting in expansion of fire susceptible woody shrubs and trees as grassland area reduced. In the case of elephants, the removal of grass by fire led to more intensive browsing pressure on trees and shrubs, leading to loss of woody vegetation. As a result, elephants and fire together tended to decrease woody vegetation cover, while hippos increased it (Lock, 1985; Lock, 1988). Culling of hippos on the Mweya Peninsula in the 1960s led to vegetation regrowth in bare patches and increased numbers of other herbivores such as buffalos (*Syncerus caffer*) and waterbuck (*Kobus defassa Uganda*). As a result, a culling program was implemented in the late 1950s and 1960s to bring the hippo numbers down in Queen Elizabeth Park (Eltringham, 1999). This program was successful in maintaining the composition of the vegetation of the park and reducing the impact of the hippos, but it was also found that this effect was temporary, as hippo reproduction increased and within 5 years the numbers were at the same level as they had been before culling.

Subsequent research in the savanna parks followed the massive crash in large herbivore numbers that occurred in the 1970s as a result of poaching by the Ugandan army and then by others after Idi Amin was driven from the country. Most of these studies have been of the behavior of particular species. Long-term projects on banded mongooses (*Mungos mungo*) (Gilchrist, 2001), lions (*Panthera leo*) (Dricuru, 1999), and lekking behavior in Uganda kob (*Kobus kob thomasi*) (Deutsch, 1994a,b,c) have been ongoing at Mweya in Queen Elizabeth and in Murchison Falls National Park. Other shorter research projects have looked at impala (*Aepyceros melampus*), bushbuck (*Tragelaphus scriptus*) (Wronski 2002, 2005), and Giant Forest Hog *Hylochoerus meinertzhageni* (H. Klingel, personal communication, 2006).

Monitoring in the savanna parks has primarily focused on populations of large mammals from aerial surveys and, to some extent, *ad hoc* monitoring of vegetation types and changes taking place as a result of the loss of large mammals.

RESEARCH IN FORESTS

Much of the research in forests has been conducted at the Makerere University Biological Field Station in Kibale National Park, by the Budongo Forest Project in Budongo Forest Reserve, and at the Institute of Tropical Forest Conservation in Bwindi Impenetrable National Park (Fig. 3.2). Initial research in Uganda's forests concentrated on developing methods to try and harvest timber sustainably. This research looked at growth rates of trees and the ecology of tropical forest types, particularly focusing on the appearance of a climax forest type dominated by *Cynometra alexandri* in medium altitude forests such as Budongo Forest Reserve. Various ways to regenerate the forest following logging were attempted: initially, seedlings were replanted but research showed that most died and the cost was too great. Other methods included the controlled shooting of elephants to stop browsing of regenerating mahogany saplings, planting of saplings, and the use of arboricide to open up the canopy and encourage more light-demanding timber species. Later research began to look at the conservation values of the tropical forests, notably the long-term studies of primates by Tom Struhsaker in Kibale Forest Reserve (Struhsaker, 1997) and the later studies of frugivory and forest regeneration by Colin and Lauren Chapman (Chapman, 1995; Chapman and Onderdonk, 1998; Chapman and Chapman, 2004). There was also research on the impacts of logging on wildlife in Kibale Forest (Skorupa, 1986; Kasenene, 1987) and Budongo Forest Reserve (Plumptre, 1996; Plumptre and Reynolds, 1994), and the recovery of forest gaps following logging (Babweteera *et al.*, 2000).

Studies of the ecology of particular species also became more common from the late 1980s with studies of chimpanzees (*Pan troglodytes*) in Kibale, Kalinzu, and Budongo forests (Hashimoto, 1995; Newton-Fisher, 2003; Reynolds, 2005; Wrangham *et al.*, 1991, 1994) and mountain gorillas (*Gorilla gorilla beringei*) in Bwindi (McNeilage *et al.*, 1998, 2006; Nkurunungi, 2003; Robbins and McNeilage, 2003), as well as studies of other primates (Plumptre, 2006; Olupot, 1999; Fairgrieve, 1995; Leland and Struhsaker, 1987). Other than a few studies of Nahna's Francolin (*Pternistis nahani*) (Sande, 2004) and some work on butterflies and fish fauna in Kibale, there had been little sustained research on the ecology of taxonomic groups other than primates. Several of these primate studies have helped develop primate tourism in Uganda during the 1990s and have contributed to improving the viewing experience for the tourists while minimizing the risks to the primate themselves (Mugisha, Chapter 11).

More recent projects in Kibale are looking at the role of disease in regulating animal populations (Gillespie and Chapman, 2006) and since the 1990s there has been a move in the direction of increased research on the impact of the parks and Forest Reserves on local communities (Hill, 1997; Naughton-Treves, 1997; Naughton-Treves *et al.*, 1998).

Monitoring work in these forests has primarily focused on primate populations and more recently on fires in Bwindi, but forest regeneration in Budongo and Kibale has also been followed.

WHERE RESEARCH HAS CHANGED MANAGEMENT STRATEGIES

We have given a very brief overview of the main lines of research that have taken place in Uganda's savannas and forests, although what we present here does not cover a fraction of the wealth of studies that exist. We have focused on those areas of research where there has been more than one study. In-depth summaries are in preparation for both savannas and forest separately and will be published in due course. During the compilation of these more detailed summaries, we attempted to compile cases where we could see research results being used to change management actions, as one of our aims was to help managers better understand their parks/reserves and assess where research results might help them better manage these areas.

Savanna research

The research in the 1950s–1970s in the savannas was much more applied in its focus than subsequent research. This was partly because of the presence of NUTAE and UIE, which facilitated longer-term applied research, whereas later research has been primarily driven by the interests of individual researchers. The early research led to changes in management strategies, including the culling of hippos in Queen Elizabeth and elephants in Murchison Falls parks and the establishment of fire management plans for each of the savanna parks. The latter promoted early burning at the start of the dry season when the fire would not burn as intensely. Research on the impact of cattle in Lake Mburo National Park led to the policy that allowed these animals into the park in the dry season both to access water and to help keep the *Acacia* woodland under control. Monitoring of the decline in large mammals in the Lake Mburo ecosystem in the 1990s led UWA to establish a pilot Sport Hunting Scheme. This scheme allowed people who owned ranches, which provided wet season forage for ungulates from the park, to benefit materially from the wildlife.

Research on invasive species has led to programs for their removal in Queen Elizabeth and Semliki National Parks. Crocodile farming was established following studies of the crocodile population at Murchison Falls National Park with a program to reintroduce a percentage of the surviving offspring to the park after 4 years. Studies of elephants and lions are leading to recommendations for regional management of these species as they move between Virunga Park in DR Congo and Queen Elizabeth. Trans-boundary collaboration and patrols along the international border have been stepped up as a result of the need to conserve species that move across international borders.

Zoning plans, which form part of the park management plans, have been developed based on the surveys of wildlife and habitat types that have occurred in these parks. A recent research project was used to stop the development of a golf course in Queen Elizabeth Park because it showed that most visitors did not want it and those that did belonged to a group who felt the fees for the park were too high at present and weren't likely to want to pay more. Speed bumps have been put along the main road through Queen Elizabeth Park to help reduce animal mortality, which monitoring had shown to be relatively high.

Forest research

Studies of forest regeneration showed that the timber species of highest commercial value often regenerated under bright light. Forest management therefore promoted arboricide treatment to remove the "undesirable" species and open up the canopy to allow in more light. This management strategy replaced enrichment planting, which was shown not to be very successful and was costly. More recent research (Plumptre, 1995) showed that many timber species don't produce fruit before they are around 50 cm diameter or greater. As a result, many were being felled before they could fruit. Management of concessions now demands that two seed trees are left in each hectare of forest, although there is little enforcement to ensure that this happens.

Research on the ecology of gorillas and chimpanzees has led to recommendations for the establishment of ape tourism sites in several forests in Uganda, particularly in Bwindi Impenetrable National Park (gorillas) and Kibale National Park, Kalinzu Forest Reserve, Kyambura Wildlife Reserve, and Budongo Forest Reserve (chimpanzees). These include rules about the number of people that can be with the apes at any one time, the minimum distance tourists can approach, and also how long they can stay. Much of the information from these studies was also used to develop Uganda's Great Ape Survival Plan, which is in the process of being implemented.

Surveys of biodiversity in these forests led to the creation of six forest parks from Forest Reserves in the early 1990s because the research had shown how globally diverse these forests were. Further survey data were used to zone the forests for conservation, minimal use, and timber harvesting as part of the Forest Department's Nature Conservation Master Plan.

In Bwindi Impenetrable National Park research on forest products used by local people has led to a legalized harvesting of selected forest products, which are monitored regularly to ensure that the harvest is sustainable. Data on illegal forest use have been used to develop snare removal programs also and studies of human–wildlife conflict have led to specific interventions such as testing Mauritius thorn *Caesalpinia decapitala* around Kibale Forest and other crops such as *Ocimum* around Budongo Forest.

Wider applications of research results

The surveys of the forests and savannas of Uganda have highlighted their global importance. As a result, many of these forests have been designated as part of the Eastern Afromontane Biodiversity Hotspot, an endemic bird area and part of the Albertine Rift Ecoregion (the ecoregion with most vertebrate species in Africa – Plumptre *et al.*, 2007).

The data on the reproduction and demography of elephants and hippos in Uganda, which were collected from the extensive culling programs, have been used elsewhere in Africa to predict population changes, and data from the long-term chimpanzee and gorilla studies are being used to compare populations across the continent.

Lessons from great ape tourism and research studies assessing the impact of tourists are being used to suggest modifications to the visitation rules and to try and ensure rules are adhered to. Studies have shown that gorillas are more stressed and feed less when the tourists are present (Muyambi, 2004), compared with just prior and just after the tourist visit. Studies of disease transmission to primates in Kibale will also be used to make better recommendations about how to reduce the risk in the near future (T. Goldberg, personal communication, 2007). The more we understand about the great apes, the more we appreciate their similarity to humans. As a result, there has been a move to try and create a new designation for great apes as World Heritage Species within UNESCO (R. Wrangham, personal communication, 2005).

SUMMARY

In both forest and savanna parks, research has been built around permanent research facilities, with most information coming from parks with research stations. Initial research in both forests and savannas was much more applied and aimed at answering management questions.

Kibale National Park is probably the leading study site in Uganda with a huge number of publications coming out of research studies conducted there compared to other research sites. Most of the research in forests is species based, looking at the ecology and behavior of individual species. Other major research in forests has focused on the effects of logging and human–wildlife conflict. In the savannas, most studies have focused on large mammals, ecological communities, and the role of fire in vegetation management. Research, both basic and applied, has contributed to management of National Parks in many ways, but it is generally difficult to draw links between specific conservation outcomes and individual projects.

Long-term research provides not only data applicable to changing situations, but also insights on threats and ways of addressing them. At the moment, much of the long-term research taking place in Uganda's Protected Areas is basic. The monitoring of large mammal species and timber plots on concessions is currently being supported by UWA and NFA respectively. Even though long-term applied research in Uganda's Protected Areas would ideally be done by the line agencies, UWA and NFA, their lack of resources for research prevents them being able to support all such projects. We believe that, if long-term applied research is to be revived in Uganda, it should be a role of these research institutions, which have a long term presence and can plan for longer-term research studies. The Directors of each research station should work closely with UWA and NFA to plan research that helps managers answer some of the key management questions that still exist. These include topics such as what drives poaching in the parks and reserves, what incentives can be created to reduce illegal activities in the parks, whether fire is changing the vegetation of the savanna parks, what methods are most effective at deterring crop raiding by various species, etc. If the research stations promoted these research topics, many of which have been identified in UWA's Research and Monitoring Plan, then the relationship between the research stations and the management institutions would become much closer and the research results would be more likely to be used by the managers as a result.

REFERENCES

Babweteera, F., Plumptre, A. J., and Obua, J. (2000). Effect of gap size and age on climber abundance and diversity in Budongo Forest Reserve, Uganda. *African Journal of Ecology*, **38**, 230–237.

Chapman, C. A. (1995). Primate seed dispersal: coevolution and conservation implications. *Evolutionary Anthropology*, **4**, 74–82.

Chapman, C. A. and Chapman, L. J. (2004). Unfavorable successional pathways and the conservation value of logged tropical forest. *Biodiversity and Conservation*, **13**, 2089–2105.

Chapman, C. A. and Onderdonk, D. A. (1998). Forests without primates: primate/plant codependency. *American Journal of Primatology*, **45**, 127–141.

Dawkins, H. C. and Philip, M. S. (1998). *Tropical Moist Forest Silviculture and Management. A History of Success and Failure*. Wallingford: CAB International.

Deutsch, J. C. (1994a). Uganda kob reproductive seasonality – optimal calving seasons or condition dependent oestrus? *African Journal of Ecology*, **32**, 283–295.

Deutsch, J. (1994b). Lekking by default – female habitat preferences and male strategies in Uganda kob. *Journal of Animal Ecology*, **63**, 101–115.

Deutsch, J. (1994c). Uganda kob mating success does not increase on larger leks. *Behavioral Ecology and Sociobiology*, **34**, 451–459.

Dricuru, M. (1999). The lions of Queen Elizabeth National Park. Makerere University Institute of Environment and Natural Resources, Kampala.
Eggeling, W. J. (1947). Observations on the ecology of the Budongo rain forest, Uganda. *Journal of Ecology*, **34**, 20–87.
Eltringham, S. K. (1999). *The Hippos. Poyser Natural History*. London: Academic Press.
Fairgrieve, C. (1995). The comparative ecology of blue monkeys (*Cercopithecus mitis stuhlmannii*) in logged and unlogged forest, Budongo Forest Reserve, Uganda: the effects of logging on habitat and population density. PhD thesis, The University of Edinburgh.
Gilchrist, J. S. (2001). Reproduction and pup care in the communal breeding banded mongoose. PhD thesis, Cambridge University.
Gillespie, T. and Chapman, C. A. (2006). Forest fragment attributes predict parasite infection dynamics in primate metapopulations. *Conservation Biology*, **20**, 441–448.
Hashimoto, C. (1995). Population census of the chimpanzees in the Kalinzu Forest, Uganda: comparison between methods with nest counts. *Primates*, **36**, 477–488.
Hill, C. M. (1997). Crop-raiding by wild vertebrates: the farmer's perspective in an agricultural community in western Uganda. *International Journal of Pest Management*, **43**, 77–84.
Kasenene, J. M. (1987). The influence of mechanized selective logging, felling intensity and gap-size on the regeneration of a tropical moist forest in the Kibale Forest Reserve, Uganda. Unpublished PhD thesis. Michigan State University, East Lansing.
Laws, R. M., Parker, I. S. C., and Johnstone, R. C. B. (1975). *Elephants and their Habitats: the Ecology of Elephants in North Bunyoro, Uganda*. Oxford: Clarendon Press.
Leland, L. and Struhsaker, T. T. (1987). Monkey business. *Animal Kingdom*, **90**, 24–37.
Lock, J. (1985). Recent changes in the vegetation of Queen Elizabeth National Park, Uganda. *African Journal of Ecology*, **23**, 63–65.
Lock, J. M. (1988). Vegetation studies in QENP. Unpublished report to Uganda Wildlife Authority, Kampala.
McNeilage, A., Plumptre, A., Brock-Doyle, A., and Vedder, A. (1998). Bwindi Impenetrable National Park, Uganda Gorilla and Large Mammal Census, 1997. Working paper No. 14. Wildlife Conservation Society.
McNeilage, A., Robbins, M. M., Gray, M. *et al.* (2006). Census of the mountain gorilla *Gorilla beringei beringei* population in Bwindi Impenetrable National Park, Uganda. *Oryx*, **40**, 419–427.
Muyambi, F. (2004). Assessment of impact of tourism on the behaviors of mountain gorillas in Bwindi Impenetrable National Park. Unpublished MSc thesis, Makerere University.
Naughton-Treves, L. (1997). Farming the forest edge: vulnerable places and people around Kibale National Park. *The Geographical Review*, **87**, 27–46.
Naughton-Treves, L., Treves, A., Chapman, C., and Wrangham, R. (1998). Temporal patterns of crop-raiding by primates: linking food availability in croplands and adjacent forest. *Journal of Applied Ecology*, **35**, 596–606.
Newton-Fisher, N. E. (2003). The home range of the Sonso community of chimpanzees from the Budongo Forest, Uganda. *African Journal of Ecology*, **41**, 150–156.
Nkurunungi, J. B. (2003). The availability and distribution of fruit and non-fruit plant resources in Bwindi: their influence on gorilla habitat use and food choice. Unpublished PhD thesis, Makerere University.
Olupot, W. (1999). Mangabey dispersal and conservation in Kibale National Park, Uganda. PhD thesis, Purdue University, West Lafayette, Indiana.
Plumptre, A. J. (1995). The importance of "seed trees" for the natural regeneration of selectively logged tropical forest. *Commonwealth Forestry Review*, **74**, 253–258.

Plumptre, A. J. (1996a). Changes following 60 years of selective timber harvesting in the Budongo Forest Reserve, Uganda. *Forest Ecology and Management*, **89**, 101–113.

Plumptre, A. J. (2006b). The diets, preferences, and overlap of the primate community in the Budongo Forest Reserve, Uganda: effects of logging on primate diets. In *Primates of Western Uganda*, ed. N. E. Newton-Fisher, H. Notman, J. D. Paterson, and V. Reynolds, Devon: Springer, pp. 345–371.

Plumptre, A. J. and Reynolds, V. (1994). The impact of selective logging on the primate populations in the Budongo Forest Reserve, Uganda. *Journal of Applied Ecology*, **31**, 631–641.

Plumptre, A. J., Davenport, T. R. B., Behangana, M. *et al.* (2007). The biodiversity of the Albertine Rift. *Biological Conservation*, **134**, 178–194.

Reynolds, V. (2005). *The Chimpanzees of the Budongo Forest: Ecology, Behavior, and Conservation*. Oxford: Oxford University Press.

Robbins, M. M. and McNeilage, A. (2003). Home range and frugivory patterns of mountain gorillas in Bwindi Impenetrable National Park, Uganda. *International Journal of Primatology*, **24**, 467–491.

Sande, E. (2004). The ecology of Nahan's Francolin, *Francolinus nahani*, in Budongo Forest Reserve, Uganda. Unpublished PhD thesis, Makerere University.

Sheil, D. (1996). The ecology of long term change in a Ugandan rain forest. Ph.D. thesis, University of Oxford.

Skorupa, J. P. (1986). Responses of rainforest primates to selective logging in Kibale Forest, Uganda: a summary report. In *Primates: The Road to Self-sustaining Populations*, ed. K. Benirschke. New York: Springer-Verlag, pp. 57–70.

Struhsaker, T. T. (1997). *Ecology of an African Rainforest: logging in Kibale and the Conflict between Conservation and Exploitation*. Florida: University Press of Florida.

Wrangham, R. W., Conklin, N. L., Chapman, C.A., and Hunt, K. D. (1991). The significance of fibrous foods for the Kibale forest chimpanzees. *Philosophical Transactions of the Royal Society of London B*, **334**, 171–178.

Wrangham, R. W., Chapman, C. A., and Chapman, L. J. (1994). Seed dispersal by forest chimpanzees in Uganda. *Journal of Tropical Ecology*, **10**, 355–368.

Wronksi, T. (2002). Feeding ecology and foraging behavior of impala in LMNP, Uganda. *African Journal of Ecology*, **40**, 205–211.

Wronski, T. (2005). Home range overlap and spatial organization as indicators for territoriality among male bushbuck (*Tragelaphus scriptus*). *Journal of Zoology, London*, **266**, 227–235.

THOMAS T. STRUHSAKER

4

Long-term research and conservation in Kibale National Park

INTRODUCTION

The need for effective conservation in Africa is urgent because of ever-increasing human pressures on Africa's forests and other ecosystems. The presence of long-term research programs can be one way to promote and support conservation.

Scientific research has been a constant presence in Kibale Forest, Uganda for nearly 40 years. From 1970–1988 I developed and managed a biological research and conservation field station in the forest. This chapter looks at the ways in which our presence in the forest contributed directly to its conservation both during those 18 years and in the 20 years since. It also examines the conservation lessons learned from my experience in Kibale and considers general strategies for conservation that can be applied to other protected areas.

Researchers working in Protected Areas, almost invariably, quickly become involved in the conservation of their field sites. Initially, the primary activity of the Kibale project was pure research on non-human primates. Within the first 2 years, however, we expanded the research to examine the effects of logging on forest regeneration, primate populations and other animal groups, and community ecology in general. The threats to the forest from poaching, timber theft, and illegal encroachment were obvious and led us to assume a far greater role in assisting the game and forest departments in protecting the forest. This assistance ranged from logistical and financial support to lobbying for National Park status for Kibale. In addition, our project also trained graduate students (Ugandans and expatriates); gave educational field trips and lectures to schoolteachers and secondary school pupils; initiated tree-planting activities with the local community; and publicized Kibale through the popular press and radio.

KIBALE NATIONAL PARK (KNP)

The Kibale National Park is a 766 km^2 ecological island of forest, colonizing bush, grassland, and swamp surrounded by a sea of agriculture dominated by tea estates and subsistence farming. The southern end of Kibale is joined to the Queen Elizabeth National Park by a 6 km wide isthmus. Kibale was originally created as a government forest and game reserve managed by the forestry department as a source of timber. Forestry agents regulated and monitored the harvest of trees and collected royalties from private timber companies that held the timber concessions and carried out the exploitation. The collection of firewood and drinking water by neighboring communities was legally permitted, but hunting and the cutting of building poles were officially prohibited. However, after the early 1970s there was little law enforcement by the forest department. To the contrary, many forest officers became involved in illegal activities (timber theft and allowing illegal agricultural encroachers into the reserve) within Kibale and all other Forest Reserves in the country (Struhsaker, 1997).

Pressure from the surrounding human population is clearly a major issue facing the Kibale Forest. These pressures are due to rapid population growth and aspirations for higher living standards, modeled after those of developed nations. Human population density around the park is relatively high. In Kabarole (previously Toro) District, where Kibale is located, the population rose from 519 821 in 1980 to 746 800 in 1991, with an annual growth rate of 3.3% and was projected to reach 944 600 by the year 2000 (NEMA, 1997). So, by the end of the year 2000, the human population density for Kabarole was at least 117 per km^2. However, in the 27 administrative parishes adjacent to Kibale National Park, the human population density was even higher; probably greater than 300 humans per km^2 in the year 2000 (NEMA, 1997).

During my studies in Kibale, human activities in the forest included poaching of duikers (*Cephalophus* spp.), bush pig (*Potamochoerus porcus*), bushbuck (*Tragelaphus scriptus*), sitatunga (*Tragelaphus spekii*), buffalo (*Syncerus caffer*), and elephant (*Loxodonta africana*); cutting of building poles and timber; production of charcoal; collection of firewood, weaving materials (reeds, palm leaves), lianas (bush rope for house construction), honey, and medicinal plants; exportation of drinking water; and harvesting of wild coffee. Monkeys and chimpanzees (*Pan troglodytes*) were rarely hunted because the predominant tribes (Batoro and Bakiga) do not usually eat primates. Occasionally, I heard of Batoro killing monkeys to eat or to feed to their hunting dogs, but this was unusual. Until the early 1960s, there were some Bakonjo living adjacent to Kibale and they did hunt primates to eat. The Bakonjo were driven out of the area in 1964 during an intertribal dispute (Struhsaker, 1975). Now that Kibale is a National Park it seems that some of these illegal activities have greatly subsided, especially elephant poaching and timber theft.

Frequent human presence in the park may lead to the introduction of exotic plants, animals, and diseases (Goldberg *et al.*, Chapter 8) that may adversely affect indigenous wildlife. With regard to plants, there is evidence suggesting that plantations of exotic species of pines that were intentionally established within Kibale when it was a Forest Reserve may have led to a localized dieback of the adults of three upper canopy tree species through the spread of a fungal pathogen (Struhsaker *et al.*, 1989). There are as yet no major exotic plant invasions of Kibale. *Lantana* sp. is found along some parts of the forest perimeter and outside of the park. *Cyphomandra betacea* (a South American Solanaceae) occurs at low densities throughout much of the forest and nowhere does it appear to exclude indigenous species. *Senna (Cassia) spectabilis* (a tropical American Caesalpiniaceae) is also widespread along forest edges and in regenerating forest.

CONSERVATION FROM RESEARCH

The Kibale project worked closely with the Uganda Game Department to assist them in protecting Kibale against hunting and the theft of timber. The department provided us with two to three armed game guards who had powers of arrest. Our role was to supervise the guards, assist with logistical support, and to occasionally patrol with them. We paid these game guards bonuses for every poacher who was arrested and convicted, for every panga (machete), spear, net, snare, gun, crosscut saw, and piece of confiscated timber they brought to us. The bonus paid for each item was about 20% below the market value in order to discourage production or purchase of these items for profit by the guards themselves. The bonus system was a highly effective incentive that allowed the guards to increase their salaries by at least two to threefold. The annual cost of this bonus system to our project in the 1980s was less than $500.

Occasionally, we sought the support of the military, which would send in 10–20 soldiers for a few days to patrol with the game guards. These activities were meant to deal with heavily armed elephant poachers, who were more numerous and better equipped than the game guards. Based on information collected by informants, the military also made visits to households and villages alleged to be harboring poachers and thieves.

We also worked closely with the judicial system. This meant regular meetings with the Resident State Attorney and Chief Magistrate in the regional and district administrative center of Fort Portal. We discussed the importance of conservation with these individuals and explained the nature of the problems in Kibale. Each time the game guards arrested poachers in Kibale, we made an effort to inform the judiciary of these arrests and to inform the police imprisoning the poachers that higher authorities had been contacted. This seemed to reduce the frequency of bribes and pre-trial releases.

With only two to three game guards, we were able to effectively protect about 80–90 km^2 of Kibale against net and snare hunters who were after duikers, pigs, bushbuck, and sitatunga. Protection was most effective in those parts of the forest where we conducted our research. We were less successful in curtailing elephant poaching, but still managed to retain greater numbers of elephants in Kibale than in any other Forest Reserve in Uganda and certainly did as well as any of the National Parks. In the 1970s and 1980s all areas of East Africa lost, on average, about 90% of their elephants, regardless of the type of economy or government. Much of this loss was probably due to corruption and participation by government officials all over Africa. What the Kibale study shows is that, even with a very small team of dedicated and honest law enforcement officers, a great deal can be achieved in terms of wildlife protection.

Throughout the 1970s we lobbied Forest Department officials and presented a strong case for research and conservation that resulted in the protection of 86 km^2 of Kibale against logging. This area was declared a combined Nature Reserve and Long-term Research Plot and closed to logging. Eventually in 1993 the entire reserve was gazetted as a National Park. The economic collapse of Uganda in the mid 1970s also had a positive impact for forest conservation because it essentially eliminated large-scale mechanized logging and forest refinement (treatment with arboricides to kill undesirable trees) that could not be maintained in the absence of foreign currency and imports of equipment, spare parts, and chemicals.

The problem of illegal agricultural encroachment in the southern part of Kibale was much more difficult for us as foreign researchers to address. This was because high-level government officials were involved in collecting bribes from the encroachers. Most of the encroachers were recent immigrants from Kabale (previously Kigezi) District known as Bakiga (descendants of the Hutu in Rwanda). Many of the encroachers admitted that they had paid bribes to the provincial forest officer, himself a Mukiga. The most blatant example of corruption involved this same forest officer who illegally appropriated a large piece of the Forest Reserve for himself and a close friend to convert to agriculture using Forest Department tractors (Struhsaker, 1997). Illegal encroachment into Kibale was further compounded by the political and social instability throughout Uganda during the 1970s and 1980s. It was not until 1992 that all illegal encroachers were removed from Kibale by President Museveni's Government.

MEASURES OF CONSERVATION SUCCESS

Duikers and elephants in Kibale

The status of plants, animals, and the ecological community as a whole is the ultimate measure of successful conservation. This requires long-term ecological monitoring and detailed studies of ecological communities, i.e., environmental audits. Some would contend that such studies and audits are too difficult and expensive to be included in conservation projects, even though financial audits are routine, mandatory, costly, and do not measure conservation success. Furthermore, alternative indices of effective conservation, such as "threat reductions" or "degree of contentment of the neighboring human population" provide no information whatever about the flora and fauna being conserved, nor are these indices necessarily less expensive to measure.

Primates are easily monitored, but they probably are not ideal for monitoring because they are not generally hunted in Kibale and so are not particularly useful in measuring the success of anti-poaching activities. Chimpanzees, however, are often caught in wire snares and in some parts of the forest as many as 20% are maimed (missing digits, hands, feet, forearms, and lower legs) (Basuta, 1987). It is not clear to what extent the chimps were caught in snares set in the forest or by those they encountered while raiding crops outside the forest.

The best indicators of conservation success at Kibale are the two most sought after prey: elephants and duikers. Thirty years ago I surveyed most of the Forest Reserves in western Uganda (Fig. 4.1) and frequently encountered elephants and their spoor in all of them. In the intervening 30 years, Kibale retained a larger population than any other Forest Reserve in Uganda and did as well as any of the National Parks (Howard, 1991). Given the similarities in habitat between these areas, the only difference between Kibale and other reserves is that Kibale benefited from some degree of law enforcement. The presence of a research team also played

Fig. 4.1. Map of Uganda showing position of the Forest Reserves in the 1970s.

a vital watchdog role in reducing elephant poaching. This was most evident following incidents of poaching when elephants would often move closer to our main research station and remain there for several days. In fact, this was sometimes our first clue that a poaching incident had occurred.

Two species of duikers (blue, *Cephalophus monticola,* and red, *C. harveyi*) were the main prey of meat hunters in Kibale. Our limited anti-poaching activities and research presence effectively protected an area of about 80–90 km^2. The research sites of Kibale had higher indices of duiker abundance than any other Forest Reserve in western Uganda (Howard, 1991) and duiker populations in the Protected Areas of Kibale may have actually increased slightly over a 20-year period (Struhsaker, 1997). There is little doubt that even limited law enforcement was instrumental in protecting large and viable populations of duikers in Kibale (also see Howard, 1991).

Other measures of conservation success

It is less clear how our efforts at training and education influenced conservation in Kibale. These are both long-term approaches to conservation and, consequently, they do not yield immediate results. Furthermore, it is often difficult to separate the effects of one strategy from others. However, the training of qualified Ugandan conservation biologists in Kibale is important in several ways. First, having Ugandan professionals working in the park increases the probability of long-term effectiveness because of the continuity and permanency they bring to the project. Second, as nationals they are often in a better position and better qualified than foreigners to deal with conflict resolution involving residents living next to the park. Three of the Ugandans who earned PhDs during my tenure in Kibale worked there extensively for the past 23–31 years as administrative and scientific directors of the Makerere University Biological Field Station at Kanyawara and project administrator for the research projects at the Ngogo camp in Kibale. This commitment by Ugandans represents an important step toward long-term conservation.

Educational field trips and discussions with secondary school teachers and their pupils probably converted some of them to conservation and certainly generated goodwill with the neighboring community. It represented an initial step toward developing attitudes more sympathetic to conservation. However, due to insufficient funds and personnel during my tenure in Kibale, not enough time and effort was invested in this strategy to yield tangible results, nor did we make any effort to evaluate the effectiveness of this approach.

CONSERVATION LESSONS: AIDS TO EFFECTIVE PARK CONSERVATION

The conservation lessons I have learned from Kibale and elsewhere are summarized in Table 4.1. These are not listed according to any rank of importance. The relative importance of each of these lessons or strategies will likely vary between sites and even vary over time within a site. Most of these points are self explanatory and obvious. However, in one way or another all of them either depend on, or are greatly facilitated by, the presence of long-term research.

Law enforcement is imperative for effective conservation, just as it is in any form of governance. However, its effectiveness is usually contingent on appropriately trained, equipped, and salaried staff. Committed leadership is critical. A bonus system for effective anti-poaching activities

Table 4.1. *Requirements for successful conservation programs in National Parks*

1.	Effective law enforcement
2.	Long-term commitment: longer than 20 years
3.	Permanent collaborative association with overseas organizations
4.	Training and participation of nationals
5.	Scientific presence and monitoring
6.	Flexible management plan: problems change with project ontogeny
7.	Education and support at both local and national levels
8.	Appropriate levels of secure funding

creates incentives for the park guards and informants. Guards must have powers of arrest. There must be an effective judicial system with checks and transparency to reduce corruption.

Long-term commitment by donors and participants is essential for success. Therefore, a secure source of funding, such as a trust fund, is crucial to sustaining project effectiveness. In order to minimize misuse of these funds, I recommend that the Principal be managed by a board of trustees outside and independent of the recipient country. Annual interest payments, on the other hand, should be managed by a board within the recipient country working with the park management authority. Annual allotments should not be excessive as they can detract from conservation objectives and lead to individual and interagency rivalries. Furthermore, annual payments should be contingent on performance, i.e., there must be accountability as determined by financial and environmental audits. Secure funding frees personnel to concentrate on conservation rather than fund raising.

More important than financial security is the long-term dedication of qualified personnel. They form the foundation of any successful conservation program. Continuity of personnel requires the existence of a professional national conservation agency that is relatively independent of shifting political winds.

One way of achieving long-term commitments is through establishing *permanent collaborative links* with overseas research and funding organizations. Such linkages help ensure continuity of the project, as well as providing a continuing source of technical expertise.

Training and participation of nationals are vital to building a core of committed personnel that is more likely to persist over the long-term than one consisting primarily of expatriates. Furthermore, nationals are more likely to remain effective during periods of political instability than are foreigners.

Scientific presence and ecological monitoring, through permanent research stations, can provide scientific information and analysis relevant to the objective management of a park.

Additionally, researchers can serve as conservation watchdogs and can provide the environmental audits that permit an evaluation of a park's conservation status. Without scientific studies and monitoring, evaluation of park effectiveness is largely subjective. Research stations should collaborate closely with park managers, but retain a high degree of autonomy in order to preserve an independent judgment.

Conservation management plans must remain flexible and sensitive to changing conditions. Long-term ecological monitoring is imperative for detecting change due to both extrinsic and intrinsic forces.

Education and public support at local and national levels serves to win the support of the people living around the park. This not only reduces external threats, but also simplifies the task of law enforcement. It is important that local populations understand why a park exists, but it is even more important to instill an ethic of respect for other species and their right to exist. When such an ethic predominates, conserving the park becomes much easier. Teachers recruited from the local community can be extremely effective in this regard. Education at the national level often relies on several different strategies including television, radio, newspapers, and lobbying politicians.

Buying the support of local populations with financial incentives is problematic at best because external market forces change and growing populations with expanding aspirations lead to ever increasing demands and expectations. Conservation can then be held hostage to the promise of escalating financial returns. The danger is that, if the benefits should decline for whatever reason, there may be no reservoir of goodwill to inhibit people from making up the deficit by extracting resources from the park. While certain economic activities such as tourism and revenue sharing can initially engage a local community in conservation, the benefits are usually too prone to the vagaries of politics and economics to have long-lasting consequences. Furthermore, in the case of Africa's forest parks, only a small proportion of the neighboring community derives financial benefits from these activities (Struhsaker *et al.*, 2005) and the realized income per capita is low. The majority of integrated conservation and development projects fall into this category. History demonstrates the tenuous nature of conservation efforts based solely on economic or other material benefits (Kramer *et al.*, 1997; Struhsaker, 1998; Oates, 1999).

A better way to relieve human pressure on a park is to develop alternative resources for neighboring residents. Privately owned wood lots are one such example. Great care must be taken to avoid development projects that attract immigration and thereby exacerbate conservation problems within the park that result from increased human populations (Oates, 1999).

The *population explosion* drives the increasing rates of consumption in the tropics and is the single most important development hampering conservation in Africa. Consequently, *family planning* education and facilities should be considered as critical components of any community conservation project. In the absence of human population stabilization, if not population reduction, we can expect a continued decline in both wildlife populations and human living standards. In those countries exporting natural resources to western industrialized nations, high levels of per capita consumption in these more developed countries also play an important role in resource depletion, especially so when these resources are grossly under-priced (Gillis and Repetto, 1988).

Beyond the local context, official support for parks and conservation in general must be gained at the national level, where crucial political and financial decisions are made.

SUMMARY

The conservation problems facing parks and other Protected Areas are multifaceted, as are the potential solutions to these problems. Because each area has its own set of peculiarities and problems, we do not expect to find a set of solutions applicable to all. However, identifying the problems and possible solutions is useful in developing realistic conservation strategies, and long-term research and monitoring are critical to this process.

ACKNOWLEDGMENTS

This chapter was adapted with permission from Struhsaker (2002). Special thanks go to Peter Marler who first suggested that I consider Kibale as a study site of red colobus. Without his encouragement in the early days of my research there, the Kibale project would not have developed nor is it likely that the park would have been created. I would also like to express my gratitude to those who helped maintain the long-term research and conservation efforts in Kibale after my departure in 1988. In particular, I would like to thank the following individuals: Isabirye Basuta, Colin and

Lauren Chapman, John Kasenene, Jerry Lwanga, John Mitani, Arthur Mugisha, David Watts, and Richard Wrangham. Elizabeth Ross is thanked for her assistance in editing this chapter.

REFERENCES

Basuta, G. M. I. (1987). The ecology and conservation status of the chimpanzee (*Pan troglodytes*) in Kibale Forest, Uganda. Unpublished PhD thesis, Makerere University, Kampala, Uganda.

Gillis, M. and Repetto, R. (1988). Conclusion: findings and policy implications. In *Public Policies and the Misuse of Forest Resources*, ed. R. Repetto and M. Gillis. Cambridge: Cambridge University Press, pp. 385–410.

Howard, P. C. (1991). *Nature Conservation in Uganda's Tropical Forest Reserves*. Cambridge, England: IUCN.

Kramer, R., Van Schaik, C., and Johnson, J. (1997). *The Last Stand*. Oxford: Oxford University Press.

NEMA (1997). National Environment Management Authority of Uganda. Kabarole District Environment Profile.

Oates, J. F. (1999). *Myth and Reality in the Rain Forest*. Berkeley and Los Angeles: University of California Press.

Struhsaker, T. T. (1975). *The Red Colobus Monkey*. Chicago: University of Chicago Press.

Struhsaker, T. T. (1997) *Ecology of an African Rain Forest: Logging in Kibale and the Conflict between Conservation and Exploitation*. Gainesville: University Press of Florida.

Struhsaker, T. T. (1998). A biologist's perspective on the role of sustainable harvest in conservation. *Conservation Biology*, **12**, 930–932.

Struhsaker, T. T. (2002). Strategies for conserving forest National Parks in Africa with a case study from Uganda. In *Making Parks Work: Strategies for Preserving Tropical Nature*, ed. J. Terborgh, C. van Schaik, L. Davenport, and M. Rao. Washington: Island Press, pp. 97–111.

Struhsaker, T. T., Kasenene, J. M., Gaither, J. C. Jr., Larsen, N., Musango, S., and Bancroft, R. (1989). Tree mortality in the Kibale Forest, Uganda: a case study of dieback in tropical rain forest adjacent to exotic conifer plantations. *Forest Ecology and Management*, **29**, 165–185.

Struhsaker, T. T., Struhsaker, P. J., and Siex, K. S. (2005). Conserving Africa's rain forests: problems in protected areas and possible solutions. *Biological Conservation*, **123**, 45–54.

NADINE LAPORTE, WAYNE WALKER,
JARED STABACH, AND FLORENCE LANDSBERG

5

Monitoring forest–savanna dynamics in Kibale National Park with satellite imagery (1989–2003): implications for the management of wildlife habitat

INTRODUCTION

Remote sensing technologies can provide detailed assessments of the state of Protected Areas, including critical information on threats (such as deforestation and wildfire), while also facilitating habitat evaluation and change detection. A multitude of satellite-based sensors of varying characteristics are now in operation, enabling the mapping of land cover and land use at various spatial and temporal scales. A new generation of high-resolution satellites, coupled with recent advances in desktop computing power and Geographic Information Systems, has greatly enhanced the ability of conservationists and park managers to integrate remote sensing information into their management plans. For example, many species are restricted to specific habitats that can now be identified with remote sensing (Turner *et al.*, 2003; Goetz *et al.*, 2007; Stickler and Southworth, in press). However, despite the advancements mentioned, the perceived complexities of remote sensing data often discourage non-specialists from leveraging this valuable resource. Here we describe how imagery acquired from the well-known Landsat class of satellites can be used to monitor long-term changes in fire regime and wildlife habitat in Kibale National Park, Uganda, thus contributing to the rich body of long-term research at Kibale focusing on conservation applications.

STUDY AREA

The Kibale National Park

Located in southwestern Uganda, the 793 km^2 Kibale National Park is one of only a few blocks of tropical forest remaining in the country. Only 3% of Uganda remains covered by rainforest, with nearly all found in the southwestern portion of the Albertine rift. Kibale itself is surrounded by a complex, fine-scale agricultural matrix where tea, coffee, bananas, corn, potatoes, and manioc fields are interspersed with monospecific plantations of Eucalyptus and pines. The park is adjacent to the northeastern border of Queen Elizabeth National Park, allowing for the migration of various species, such as elephant, between the two Protected Areas. Kibale is known for its remarkable diversity of 11 primate species, which include the black and white and the red colobus monkey, L'Hoest monkey, and the eastern chimpanzee. In addition, red and blue duikers, giant forest hogs, golden cats, civets, the Congo clawless otter, and 335 species of birds have been sighted (Howard, 1991).

The terrain within Kibale is characterized by gentle hills spanning the dominant north to south elevation gradient, ranging from a high point of 1590 m in the northern portion of the park to 1110 m near the southern boundary. Average annual rainfall over the park is 1750 mm, which is concentrated in two peak periods, one in April and another in October. The average daily temperature ranges from a minimum of 14 °C to a maximum of 27 °C.

The vegetation of Kibale National Park can be divided into six general categories: forests (58%), grasslands (15%), abandoned farms (10%), degraded forest (9%), woodland (6%), and wetlands (2%) (Lwanga *et al.*, 2000). Wetlands are found within the lower-lying floodplains surrounding the two large rivers draining the park, the Mpanga and the Dura. Each flows south and empties into Lake George. These swampy areas are typically dominated by stands of papyrus and various species of palm. Forests in the northern portion of the park, where the elevation is typically above 1500 m, are dominated by evergreen assemblages and are generally rich in timber species. The dominant canopy species is the mubura (*Parinari excelsa*); sub-dominant species include *Carapa grandiflora*, *Aningeria altissma*, and *Newtonia buchananii* (Howard, 1991). The southern portion of Kibale is characterized by a greater proportion of deciduous tree species than in the north. Forests here are dominated by *Olea welwitschii*, *Pterygota mildbraedi*, and *Cynometra alexandri*. Forests in the central portion of the park are composed of a mix of species that characterize the gradient between evergreen and semi-deciduous mid-altitude moist tropical forest. In general, the forest understory in all regions of the park is sparse due to the dense canopy cover; denser understory may be found in areas that have experienced some degree of canopy disturbance (e.g., logging and swampy valley bottoms). Grasslands are found generally on hilltops in the north and throughout major portions of the south. The origins of many of these non-wooded areas can be traced back to human settlement (Lwanga, 2006), where fires were historically used by hunters to maintain grasslands for the purpose of attracting game.

Historic land use and land management policies

Prior to being formally gazetted as a National Park in 1993, Kibale had been a Forest Reserve since 1926, providing large volumes of timber (Howard, 1991). The northern portion of the park was targeted in particular for its valuable timber species, including *Cordia millenii* and *Entandrophragma angolense*, resulting in changes in forest structure and function that continue to impact the ecosystem today (Struhsaker, 1997; Chapman *et al.*, 2000a). The Forest Reserve was contiguous with the former Kibale/Queen Elizabeth Game Corridor, of which 134 km^2 had dual status. Van Orsdol (1986) estimated that 640–960 households were living in the southern part of the Forest Reserve in 1982, translating into an area of approximately 18–27 km^2 of active encroachment, based on an average household cultivation area of 2.8 ha. It was not until the 1990s that the Ugandan Government removed all people living within the borders of the Protected Area. As these individuals were nearly all involved in some form of small-holder agriculture, the impact of their eviction on land cover/use within the occupied portions of the park was significant (Chapman and Lambert, 2000). Annual burning, commonly used by farmers during the dry season to clear debris and encourage crop growth, would have largely ceased. Consequently, the eviction of farmers and the resulting abandonment of farmland reduced fire occurrence and permitted the recolonization of open lands by trees and shrubs.

In 1994, a carbon sequestration project was launched in the southern part of the park (FACE, 2007). The project focused on the restoration of natural forest using native tree species, and fire control. Replanting activities will be implemented over a 17-year period, with a project lifetime of 99 years. The average carbon benefit is estimated to be 26 t C ha^{-1}, at 0.9 t C ha^{-1} yr, with a total of 707 000 t C over the project lifetime (Watson *et al.*, 2001).

It is well known that fire is the single most important factor limiting the recolonization of grasslands by trees and shrubs (Lwanga, 2003;

Nangendo *et al.*, 2006). In the Budongo Forest Reserve, located north of Kibale National Park, fire suppression has facilitated the invasion of fire-intolerant species, and this recolonization is taking place without re-planting. While visiting Kibale in 2006, little difference was observed between the replanted and naturally regenerating areas, suggesting that fire suppression alone was the most important factor for recolonization. In the continued absence of fire, forest cover will re-establish itself as forest tree species gain a foothold in grasslands, canopy closure increases, and groundcover vegetation and litter become too moist to propagate fire (Nangendo *et al.*, 2006). In addition, fires can maintain medium levels of disturbance to ecosystems (i.e., the intermediate disturbance hypothesis), creating a more complex ecotone between grasslands and forest ecosystems that produces a more compositionally and structurally diverse landscape, which is likely to support higher biodiversity (Grime, 1973; Connell, 1978).

DATA SETS AND METHODS

Land cover change analysis

A change-detection analysis was conducted using Landsat data from two separate acquisitions in 1989 and 2003 (Fig. 5.1). Both the 1989 and 2003 Landsat scenes were registered to an orthorectified Landsat scene from 2001 (Tucker *et al.*, 2004) using the nearest-neighbor resampling method at a resolution of 28.5 m. Georeferencing was based on a second-order polynomial transformation (RMSE = 18.6 m) using 35 ground control points (GCPs) for the 2003 image. GCPs were identified throughout the Landsat scene with greater emphasis placed on regions within and surrounding Kibale National Park. The 1989 scene was similarly registered to the georeferenced 2003 scene using a first-order polynomial transformation (RMSE = 17.0 m) based on 21 GCPs.

The change-detection analysis was performed using eCognition 4.0 with Landsat bands 2, 3, and 4 of the 1989 and 2003 images serving as inputs. The analysis consisted of two principal steps. First, a segmentation was conducted on the six-band image stack. Here, the term "segmentation" refers to the subdivision of an image (or image stack) into a number of regions (i.e., polygons), based on some predefined criteria (Baatz *et al.*, 2004). The eCognition segmentation algorithm is a bottom-up, region-merging technique that starts with single-pixel regions commonly referred to as "image objects." Larger objects are formed through a pairwise clustering process. A scale factor of 10 was selected, corresponding to the intended average size (in pixels) of image objects produced by the segmentation.

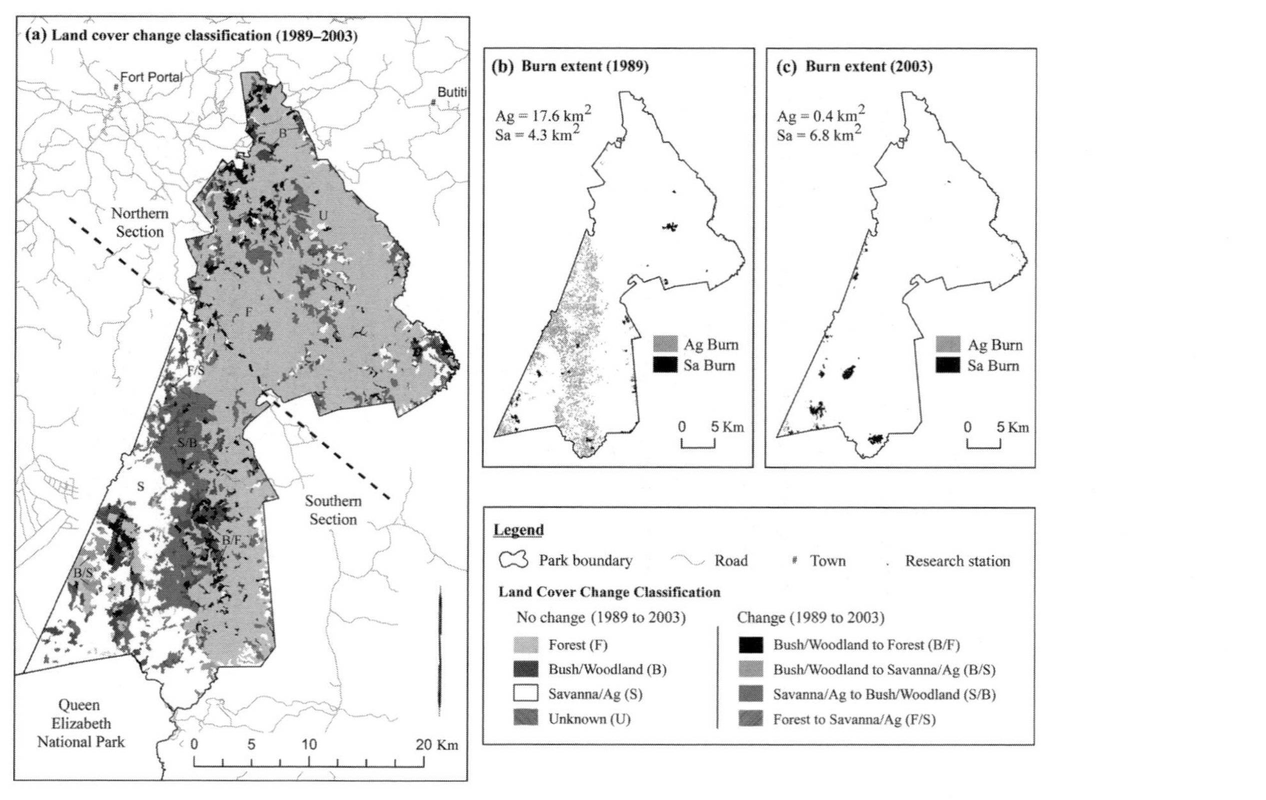

Fig. 5.1. (a) Kibale National Park land cover change classification. The park was arbitrarily separated into two sections for statistical comparisons (see Table 5.2), denoted by the dashed line. (b and c) Analysis of total area burned, separated into Agriculture (Ag) and Savanna (Sa) burns. Data projected to UTM 36N, Arc1960 datum. Color version available (ftp://whrc.org/Africa/Albertine/Kibale).

In the second step, the resulting image objects were subject to a standard nearest-neighbor supervised classification routine provided within eCognition. To train the nearest-neighbor classifier, several image objects (~10–20) were identified in six of nine possible change-detection classes represented by transitions among forest, bushland/woodland, and savanna/agriculture classes (Fig. 5.1). The classification routine was run using the following object-based metrics as inputs: (1) object means computed from each of the six Landsat bands, (2) texture after Haralick (GLCM mean and standard deviation for all directions (Baatz *et al.*, 2004)), (3) Normalized Difference Vegetation Index (NDVI) based on both the 1989 and 2003 imagery (Tucker, 1979).

Following classification, the change-detection product was evaluated visually against the image inputs as well as against the authors' own on-the-ground experience. Modifications to the final map were made where obvious errors were identified. Additional validation was conducted utilizing a data set provided by the Wildlife Conservation Society (Plumptre *et al.*, 2003). These data were collected sporadically from March 12, 2000 until June 16, 2000 to assess the status of wildlife populations throughout the park. Data consisted of 2086 GPS data points that were collected along line transects. At each data point, a characterization of land cover was made. However, since most of the data points (1926) were collected in forest (Closed Tropical High Forest, Open Tropical High Forest, and Eucalyptus plantations), only the accuracy of pixels that were classified as "forest" in the image were assessed.

Burned area change analysis

Following the eviction of small-holder farmers, an investigation was undertaken to identify and quantify the area of burn scars in each Landsat image (i.e., 1989 and 2003). Both agricultural and savanna burns were identified. Agriculture burns are generally small in size and associated with the preparation of agricultural fields, while savanna burns are larger and typically associated with hunting or cattle husbandry, with the intent of the burning being to encourage the new growth of grasses (Malingreau *et al.*, 1990).

The analysis was performed using eCognition 4.0 with Landsat bands 3, 4, and 5 of the 1989 and 2003 images serving as inputs. The analysis consisted of two principal steps. First, a segmentation was conducted on each three-band image stack. A scale factor of 2 was selected, corresponding to the intended average size (in pixels) of image objects produced by the segmentation. In the second step, a simple thresholding technique was used to classify the burn scars in each image. A thresholding approach is well suited to burn-scar classification, given that burn scars tend to be the darkest features in the imagery. Thresholding was conducted using the following object-based metric as an input: ratio of the image–object mean values in band 4 divided by the sum of the image–object mean values in all bands (3, 4, 5). Following the burn scar classification, burn scars were further identified using visual interpretation (i.e., on-screen) as either (1) agriculture burns (i.e., burns associated with small-holder agriculture) or (2) savanna burns associated with larger existing savannas/bushlands/woodlands. Area statistics were subsequently generated for burn scars identified in each image (1989/2003) and for each burn scar type (agriculture/savanna).

RESULTS

The results of the land cover change analysis indicate that the area under forest cover within Kibale National Park was quite stable over the 14-year period of study (Fig. 5.1(a)), with forest loss (i.e., deforestation) limited to less than 6 km^2 (~1%) of the total park area (Table 5.1). Overall, 53% of the park (~420 km^2) remained as forest between the 1989 and 2003 time periods (Table 5.1).

In comparison, significant changes were observed in the area covered by savanna/agriculture between the two time periods, with nearly 103 km^2 (~ 13% of the park) shifting from the savanna/agriculture class to the bushland/woodland class. By 2003, an additional 4% of the park (~ 34 km^2) had changed from the bushland/woodland class to the forest class. Thus, in total, over 17% (~ 137 km^2) of the park experienced a loss of grasslands or otherwise non-forested cover as a result of colonization by tree and shrub species from adjacent forested areas (Table 5.1).

Overall, the most striking changes were observed in the southern portion of the park where fire suppression combined with reforestation (i.e., tree planting and/natural regeneration) led to forest recolonization. In the south (see Fig. 5.1), approximately 27% of the total park area was mapped as savanna/agriculture in 1989 but only 19% mapped as such in 2003 (Table 5.2). In the north, where the overall extent of savanna/agriculture is considerably less than in the south (49.9 km^2 compared to 212.5 km^2 in 1989), the total extent of savanna/agriculture decreased from roughly 6% of the total park area in 1989 to just under 3% in 2003 (Table 5.2). The results of the burn scar mapping exercise reveal clear differences in burn scar area, density, and distribution across the park. In 1989, a significant portion (80%) of the total area burned (22 km^2) was associated with agriculture (Fig. 5.1(b)). By 2003, the total area burned had decreased markedly to

Table 5.1. *Land cover change classification summary statistics*

	Forest	Bushland/woodland	Savanna/agriculture	Unknown (cloud)	1989 Totals
Forest	**420.1 (53.0%)**	–	*5.5 (0.7%)*	–	425.5 (53.7%)
Bushland/woodland	*34.4 (4.3%)*	**62.4 (7.9%)**	*7.5 (1.0%)*	–	104.3 (13.2%)
Savanna/agriculture	–	*102.9 (13.0%)*	**159.6 (20.1%)**	–	262.5 (33.1%)
Unknown (cloud)	–	–	–	0.5 (0.1%)	**0.5 (0.1%)**
2003 Totals	454.4 (57.3%)	165.3 (20.9%)	172.5 (21.8%)	0.5 (0.1%)	792.7 (100.0%)

Note: Classification compares Landsat TM imagery acquired September 21, 1989 with Landsat ETM+ imagery acquired January 31, 2003. Table details the total area (km^2) unchanged between the two time periods (bold) while also indicating the area that transitioned to different land cover types (italic). Row totals summarize the total area in each land cover class in 1989; column totals summarize the area in each class in 2003.

Table 5.2. *Land cover change analysis comparing the northern and southern sections of Kibale National Park (792.7 km^2)*

	Northern section (380.3 km^2)				Southern section (412.4 km^2)			
	1989		2003		1989		2003	
	Area (km^2)	%	Area (km^2)	%	Area (km^2)	%	Area (km^2)	%
Forest	288.7	36.4	304.6	38.4	136.8	17.3	149.9	18.9
Bushland/woodland	41.5	5.2	54.6	6.9	62.8	7.9	110.8	14.0
Savanna/agriculture	49.9	6.3	21.0	2.6	212.5	26.8	151.5	19.1
Unknown (cloud)	0.2	0.0	0.2	0.0	0.3	0.0	0.3	0.0

Note: A Landsat TM satellite image acquired September 21, 1989 was used as the reference image and a Landsat ETM+ satellite image acquired January 31, 2003 was compared to it. The "Forest" class consists of both natural forest and tree plantations. Percent (%) was calculated with respect to the total park area.

7 km^2, with the majority of the area (94%) associated with savanna rather than with agriculture burns (Fig. 5.1c). During both periods, the vast majority of burns (both agriculture and savanna) were located in the southern portion of the park (Fig. 5.1(b), (c)).

DISCUSSION

While previous research in Kibale National Park has focused on the loss or degradation of primate habitat resulting from the expansion of logging and/or agriculture (Kasenene, 1987; Struhsaker *et al.*, 1996; Struhsaker, 1997; Chapman and Lambert, 2000), this research suggests that the cessation of small-holder agriculture, most notably in the southern portion of the park, combined with fire suppression policies throughout the park, has led to the loss of grassland and other open lands and the establishment of colonizing forest (~ 34 km^2) and a mixed formation of bushland and woodland (~ 103 km^2). This colonization is certainly facilitated by the proximity of old growth forest, the presence of seed dispersers, and the policy of park-wide fire suppression enforced by the Uganda Wildlife Authority.

The 1989 estimate of burned area associated with farming activities (~18 km^2) corresponds well with those of Van Ordsol (1986), who documented 18–27 km^2 of active encroachment in 1982. The subsequent decline in area burned and the concomitant and ongoing change in vegetation composition and structure is likely to impact the distribution and density of wildlife species throughout the park. Recent work by Lwanga (2006) suggests that secondary forest has the potential to support a number of primate species including black and white colobus (*Colobus guereza*), redtail (*Cercopithecus ascanius*), and possibly blue monkeys (*Cercopithecus mitis*). Black and white colobus were encountered more frequently in colonizing forests than in old-growth forests but the encounter rate was lower for chimpanzees (*Pan troglodytes*).

Using Chapman and Lambert's (2000) density estimates of 665.9 monkeys and 2.53 chimpanzees per square kilometer of forest in Kibale National Park, we might expect a gross increase in the primate population of approximately 91 228 monkeys and 347 chimpanzees. This calculation is based on the assumption that the 137 km^2 of bushland/woodland and new forest will become suitable primate habitat within the next 20 years. Obviously, for some primate species such as the black and white colobus, the forest is already suitable (Lwanga, 2006). Lwanga (2003) argues that small bodied primates, such as redtails, may prefer areas of recolonizing forest because these serve as regular sources of year-round fruit. Early successional plant species begin fruiting at an early age and fruit for long periods (Whitmore, 1998), making areas with these species particularly popular during periods when fruit availability in the primary forest is low. Additionally, arthropod populations are higher in early successional forests and represent 22% of the redtail diet (Struhsaker, 1997). Chapman *et al.* (2000b) also documented that redtail, blue, red colobus, and black-and-white colobus declined appreciably in unlogged primary forests. Areas colonized by early successional species might provide alternative and perhaps more attractive habitat for these primates. Studies of red colobus in Kibale have found this species to be particularly flexible in terms of both plant species and plant parts exploited for food (Struhsaker, 1978; Chapman *et al.*, 2000b). This flexibility in dietary requirements might explain in part why red colobus can be found in colonizing forests.

In Costa Rica, Fedigan and Jack (2001) found that howler (*Allouatta palliata*) and capuchin (*Cebus capucinus*) monkey populations recover only 28 years after the onset of fire suppression and active management of abandoned cattle ranches. Non-specialized primate feeders are known to survive in various habitats including degraded forest, and these areas should not be disregarded as worthless for conservation but should be protected against fire and other degrading activities to allow recolonization by forest and forest animals. Remote sensing does allow us to quickly assess forest succession but tells us nothing about factors that influence such changes in vegetation cover. In areas with a history of fire, such as the southern section of Kibale, the contribution of the soil seed bank to forest succession may be minimal or non-existent, suggesting that seed dispersing agents play a vital role in forest succession. At Ngogo, Kibale National Park, Lwanga (2003) found that animal-dispersed tree species in secondary forests where fire had been suppressed for 25–30+ years were significantly more abundant than non-animal-dispersed species. Hornbills, civets, and large frugivores such as chimpanzees and baboons are all known to be important seed dispersers in Kibale National Park (Howard, 1991). Additionally, elephants in Kibale and probably elsewhere are the only dispersers of the large seeds (88 mm long) of *Balanites wilsoniana* (Chapman *et al.*, 1992). Consequently, the role of animals in facilitating forest recolonization of the southern grasslands of Kibale cannot be underestimated and should be evaluated in relation to changes in fire and wildlife management.

SUMMARY

In southwestern Uganda, the suppression of fire in portions of Protected Areas that were once subject to regular burning has greatly impacted on forest/savanna dynamics. These changes are likely to have important ramifications for the densities and distributions of wildlife populations. In the northern section of Kibale, where forest cover dominates the landscape, a management strategy based on periodic prescribed fire will be necessary if the area remaining in savanna (< 3% of the park; Table 5.2), and the associated forest/savanna interface, is to be maintained. In the absence of fire, continued forest colonization of these savannas will surely benefit forest primates. However, the continued loss of grasslands is also likely to negatively affect the survival of savanna-dependent wildlife. Overall, maintaining both forest and savanna ecosystems in Kibale National Park will create a more compositionally and structurally diverse landscape, which is likely to support higher biodiversity.

As illustrated here, remote sensing technologies are ideally suited to aid conservationists and park managers in quantifying the impact of management policies on habitat dynamics. When integrated with data from long-term field studies, these tools allow for assessments that are both spatially and temporally synoptic, enhancing our ability to monitor species, populations, and ecosystems at the landscape level and beyond.

ACKNOWLEDGMENTS

This research was supported by the NASA Applied Sciences Program, grant NNS06AA06A. We thank J. Lwanga, A. Plumptre, R. Wrangham, E. Ross, S. Goetz, and A. White for their comments on the manuscript.

REFERENCES

Baatz, M., Benz, U., Dehghani, S. *et al.* (2004). eCognition: user guide 4. Definiens Imaging GmbH, pp. 485.

Chapman, C. A. and Lambert, J. E. (2000). Habitat alteration and the conservation of African primates: case study of Kibale National Park, Uganda. *American Journal of Primatology*, **50**, 169–185.

Chapman, L. J., Chapman, C. A., and Wrangham, R. W. (1992). *Balanites wilsoniana: elephant dependent dispersal? Journal of Tropical Ecology*, **8**, 275–283.

Chapman, C. A., Balcomb, S. R., Gillespie, T. R., Skorupa, J. P., and Struhsaker, T. T. (2000a). Long-term effects of logging on African primate communities: a 28-year comparison from Kibale National Park, Uganda. *Conservation Biology*, **14**, 207–217.

Chapman, C. A., Chapman, L. J., and Gillespie, T. R. (2000b). Scale issues in the study of primate foraging: red colobus of Kibale National Park. *American Journal of Physical Anthropology*, **117**, 349–363.

Connell, J. H. (1978). Diversity in tropical rain forests and coral reefs. *Science*, **199**, 1302–1310.

Fedigan, L. M. and Jack, K. (2001). Neotropical primates in a regenerating Costa Rican dry forest: a comparison of howler and capuchin population patterns. *International Journal of Primatology*, **22**, 689–713.

FACE (2007). The Face Foundation. http://www.stichtingface.nl/.

Goetz, S., Steinberg, D., Dubayah, R., and Blair, B. (2007). Laser remote sensing of canopy habitat heterogeneity as a predictor of bird species richness in an eastern temperate forest, USA. *Remote Sensing of Environment*, **108**, 254–263.

Grime, J. P. (1973). Competitive exclusion in herbaceous vegetation. *Nature*, **242**, 344–347.

Howard, P. C. (1991). Nature conservation in Uganda's tropical forest reserves. Gland, Switzerland: IUCN, pp. 313.

Kasenene, J. M. (1987). The influence of mechanized selective logging, felling intensity, and gap-size on the regeneration of moist tropical forest in Kibale Forest Reserve, Uganda. Unpublished PhD thesis, Michigan State University, East Lansing.

Lwanga, J. S. (2003). Forest succession in Kibale National Park, Uganda: implications for forest restoration and management. *African Journal of Ecology*, **41**, 9–22.

Lwanga, J. S. (2006). Spatial distribution of primates in a mosaic of colonizing and old growth forest at Ngogo, Kibale National Park, Uganda. *Primates*, **47**, 230–238.

Lwanga, J. S., Butynski, T. M., and Struhsaker, T. T. (2000). Tree population dynamics in Kibale National Park, Uganda 1975–1998. *African Journal of Ecology*, **38**, 238–247.

Malingreau, J. P., LaPorte, N., and Gregoire, J. M. (1990). Exceptional fire events in the tropics – southern Guinea, January 1987. *International Journal of Remote Sensing*, **11**, 2121–2123.

Nangendo, G., Steege, H., and Bongers, F. (2006). Composition of woody species in a dynamic forest–woodland–savanna mosaic in Uganda: implications for conservation and management. *Biodiversity and Conservation*, **15**, 1467–1495.

Plumptre, A. J., Behangana, M., Davenport, T. *et al.* (2003). The biodiversity of the Albertine rift. Albertine Rift Technical Report, No. 3, pp. 105.

Stickler, C. M. and Southworth, J. (in press). Application of multi-scale spatial and spectral analysis for predicting primate occurrence and habitat associations in Kibale National Park, Uganda. *Remote Sensing of Environment*.

Struhsaker, T. T. (1978). Food habits of five monkey species in the Kibale Forest, Uganda. In *Recent Advances in Primatology*, vol. 1, *Behaviour*, ed. D. J. Chivers and J. Herbert. London: Academic Press, pp. 225–248.

Struhsaker, T. T. (1997). *Ecology of an African Rain Forest: Logging in Kibale and the Conflict between Conservation and Exploitation*. Gainsville, Florida: University Press of Florida, pp. 432.

Struhsaker, T. T., Lwanga, J. S., and Kasenene, J. M. (1996). Elephants, selective logging and forest regeneration in the Kibale Forest, Uganda. *Journal of Tropical Ecology*, **12**, 45–64.

Tucker, C. J. (1979). Red and photographic infrared linear combinations for monitoring vegetation. *Remote Sensing of Environment*, **8**, 127–150.

Tucker, C. J., Grant, D. M., and Dykstra, J. D. (2004). NASA's global orthorectified landsat data set. *Photogrammetric Engineering and Remote Sensing*, **70**, 313–322.

Turner, W., Spector, S., Gardiner, N., Fladeland, M., Sterling, E., and Steininger, M. (2003). Remote sensing for biodiversity science and conservation. *Trends in Ecology and Evolution*, **18**, 306–314.

Van Orsdol, K. (1986). Agricultural encroachment in Uganda's Kibale Forest. *Oryx*, **22**, 115–118.

Watson, R., Noble, I. R., Bolin, B., Ravindranath, N. H., Verardo, D. J., and Dokken, D. J. (2001). *Land Use, Land-use Change, and Forestry*. Cambridge: Cambridge University Press (http://www.ipcc.ch/pub/reports.htm), pp. 375.

Whitmore, T. C. (1998). *An Introduction to Tropical Rain Forests*, 2nd edn. Oxford.: Oxford University Press.

COLIN A. CHAPMAN, LAUREN J. CHAPMAN,
PATRICK A. OMEJA, AND DENNIS TWINOMUGISHA

6

Long-term studies reveal the conservation potential for integrating habitat restoration and animal nutrition

INTRODUCTION

Human modification of ecosystems is threatening biodiversity on a global scale. The net change in global forest area between 2000 and 2005 was $\approx$ 7.3 million ha per year ($\approx$200 km^2 of forest per day) (FAO, 2005). This does not consider the vast areas being logged selectively or the forests degraded by fire, both of which can impact huge areas. For example, during the 1997/1998 El Niño, 7 million ha of forest burned in Brazil and Indonesia alone (Chapman and Peres, 2001). And, even when the physical structure of the forest remains intact, subsistence and commercial hunting can have a profound impact on forest animal populations. For example, Chapman and Peres (2001) estimate that 3.8 million primates are consumed annually in the Brazilian Amazon.

In Uganda, the country where this study focuses, threats to biodiversity are similarly grave. Closed-canopy tropical forest once covered 20% of the country's land area, but deforestation has reduced this to just 3% (Howard *et al.*, 2000). Furthermore, Uganda lost 18% of its remaining forest between 1990 and 2000 (Howard *et al.*, 2000). The most recent estimate suggests that the annual rate of loss of tropical high forest is 7%, while that of woodland is 5% and bushland is 4% (Pomeroy and Tushabe, 2004).

One conservation strategy for addressing these pressures on tropical ecosystems is to protect specific areas in as pristine condition as possible; namely in parks and reserves. However, these Protected Areas are often small in area, and it is widely acknowledged that not all biodiversity will be maintained in such isolated, small habitats. Thus, associated with this strategy, it is anticipated that, once economic and social situations in developing countries have changed, these areas will serves as the seeds of recovery, particularly for organisms requiring resources beyond the confines of the park boundary. For this strategy to be viable requires a number of conditions to be met. First, the integrity of Protected Areas must be maintained over the time frame needed for economic and social change to occur. This is certainly a difficult task and the duration over which this change will occur is often not known. Second, the areas outside of Protected Areas must not be degraded to such an extent that recovery is arrested or does not occur on a time frame that is suitable for management. A recent review of the accumulation of vegetative biomass on degraded tropical lands suggests that the land-use practices typically used in the tropics result in lands with sufficient resources to promote regeneration in a time frame that is suitable for management (Naughton-Treves and Chapman, 2002). Unfortunately, the few available studies from Uganda suggest that rates of tree regeneration here are amongst the slowest in the world, with regeneration often being suppressed by grasses and herbs (*Acanthus pubescens*) (Naughton-Treves and Chapman, 2002; Paul *et al.*, 2004). Third, and possibly most critically, the scientific community must have a working knowledge of how to restore specific plant and animal communities. This requires both long-term studies that monitor population changes over biologically meaningful time frames and the knowledge of how to integrate information on habitat restoration and determinants of animal abundance. Unfortunately, restoration ecology historically has been biased to investigations of processes affecting plant communities. To verify this, we reviewed all the research articles published in the journal *Restoration Ecology* from 2000 to the last issue of 2006 ($n = 372$ articles; excluding invited issues). Of those articles published in *Restoration Ecology* only 11.8% involved animals or an animal-mediated process (e.g., seed dispersal, pollination). Only 1.6% of the articles published in this journal involved descriptions of the recovery or restoration of animal populations (there was a special issue on this topic in 2001 with six articles).

Given the paucity of studies that deal with restoration of animal populations, the lack of studies integrating determinants of animal abundance and restoration ecology, and the documented slow regeneration in Uganda, we initiated a series of studies in Kibale National Park, Uganda to determine the critical factors that influence the abundance of endangered red colobus monkeys (*Procolobus rufomitratus*) and sought to understand how to facilitate forest regeneration in such a way as to promote a forest that would support high red colobus abundance.

DETERMINANTS OF RED COLOBUS ABUNDANCE

As we came to know Kibale, we became intrigued by the apparent variation in primate abundance, and realized that this variation offered a unique opportunity to investigate the ecological determinants of monkey abundance. To determine the extent of variation in red colobus density, we conducted intensive line transect surveys at six sites typically every second week for 2 years (Chapman and Chapman, 1999). To establish which food resources were important, we collected more than 1000 hours of feeding observations and determined the abundance of the major food resources at each of the six sites to evaluate whether red colobus abundance was related to food availability. We found that red colobus numbers were fairly high at most sites, even in disturbed areas. However, a surprisingly low population density was found at Dura River, a relatively undisturbed riverine site in the middle of the park. As we predicted, red colobus density was related significantly to the cumulative size of important food trees, but only when the Dura River site was excluded.

We initially thought the red colobus monkeys were below carrying capacity at Dura River. A small number of censuses conducted in 1970 and 1971 (Struhsaker, 1975, 1997) estimated red colobus group density to be 2.7 times greater than what we recorded in 1996–1997. An epidemic reportedly had killed a number of male red colobus monkeys in another area of Kibale in the early 1980s (T. T. Struhsaker personal communication). If such epidemics are common, or if hunting by chimpanzees (*Pan troglodytes*) at this site had recently been intense, the colobus population could be below carrying capacity. However, another option to consider is the possibility of variation in food quality that may not be represented in our estimates of food availability.

Unlike most primates, colobus monkeys have a specialized alkaline fore-stomach designed for digesting high-fiber leaf material. Milton (1979) proposed that the ratio of protein to fiber in food items was a good predictor of food quality, because it reflects both nutritional value and digestibility. By using the ratio of protein to fiber as an index of mature leaf quality, several subsequent studies found positive correlations between colobine biomass and food quality at local (Ganzhorn, 2002) and regional (Oates *et al.*, 1990) scales. To apply this model, we quantified the degree to which the average protein to fiber ratio of mature leaves at a site could predict the biomass of red and black-and-white colobus at four sites in Kibale. Although our sample size was too small for robust statistical analyses, our results suggested that colobus biomass was related positively to the average protein to fiber ratio of mature leaves across sites. Most remarkably, when we accounted for food quality in this manner, the low population density at Dura River was no longer an anomalous outlier. It thus appears that, although food is abundant at Dura River, it is of low quality, and this is likely the reason that the site does not support a large colobus population.

While these studies suggest that the protein to fiber ratio of available foods may limit the size of folivore populations, we felt that the data were insufficient to convince managers to use these principles in constructing management plans. We felt that we needed several independent populations to increase our sample size and to develop a more rigorous predictive model. To do this, we turned to a series of forest fragments outside of Kibale. These forest fragments varied in size and composition, and provide a quasi-experimental setting that allowed us to investigate the influence of this ecological variation on primate populations. Before making any comparisons across fragments, we wanted to establish which populations were stable. If some populations were not at carrying capacity due to recent effects of disease, habitat loss, or hunting, then correlations between food availability and/or quality and folivore biomass could be spurious. In 1995, we surveyed the primate communities in 20 of these forest fragments to determine the abundance of black-and-white colobus monkeys. In 2000 and 2003, we resurveyed these fragments to assess population and forest stability, and to compare monkey biomass to the protein to fiber ratio of leaves in those fragments that we determined to have stable populations (Chapman *et al.*, 2004).

We discovered that 3 of the 20 fragments inhabited by primates in 1995 had been cleared, and resident primate populations were no longer present. These fragments had remained intact since at least the 1940s, but recent economic conditions had led to more rapid deforestation. Most fragments had been cleared for charcoal production, gin-brewing, brick-making, or timber extraction. In the remaining fragments, the total black-and-white colobus populations had declined by 40% in just 5 years. While we had initially hoped that most colobus populations in the fragments would be stable, we found that there were only five stable populations. Although this was alarming from a conservation perspective, these five sites increased our sample size sufficiently to conduct a more robust statistical analysis of the protein to fiber model. Across these five fragments, colobus biomass was correlated with the protein to fiber ratio ($r^2 = 0.730$, $P = 0.033$). To examine the model more rigorously, we

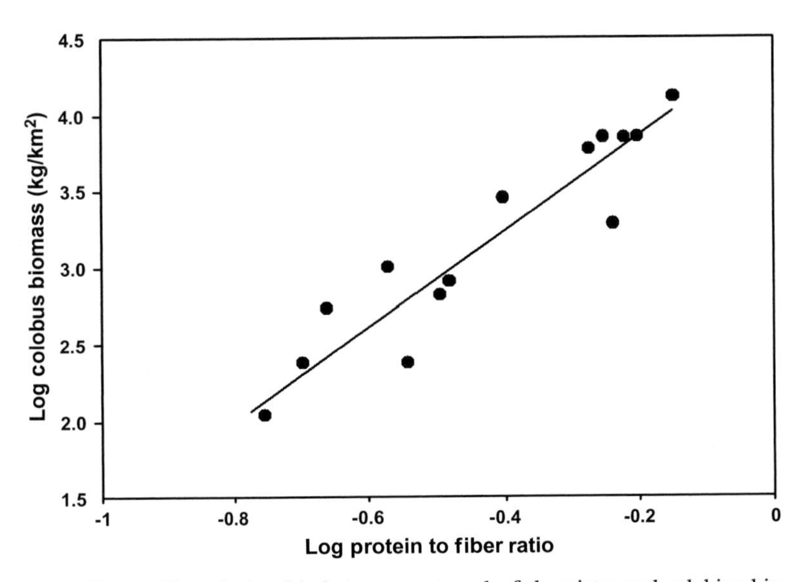

Fig. 6.1. The relationship between mature leaf chemistry and colobine biomass at rainforest sites in Africa and Asia.

combined the data from the fragments with the four sites from within Kibale and five published values from other sites in Africa and Asia. Colobine biomass across all 14 sites could be predicted with a significant level of confidence from the protein to fiber ratios of available mature leaves ($r^2 = 0.869$, $P < 0.001$; Fig. 6.1).

While the mechanism behind this correlation is not well understood, these results suggest that, if managers could provide colobus with foods high in protein and low in fiber, their populations would flourish. Studies such as this provide managers with alternative approaches for conservation. In the past, managers typically have promoted an animal species by removing threats; however, now the opportunity exists to plan to augment populations by providing high quality foods.

RESTORATION OF A PLANT COMMUNITY CONDUCIVE TO PROMOTING RED COLOBUS ABUNDANCE

This research only addressed half of what we wished to understand; we still needed to gain an understanding of how we could facilitate forest regeneration in such a way as to promote a forest that would support high red colobus abundance. An understanding of forest restoration processes could be very important for Uganda and Kibale, in particular. First, as economic situations change in Uganda and other African countries, we feel that there will be increasing interest in forest restoration. This may be encouraged by repercussions of the Kyoto Protocol. When industrialized countries fail to meet their commitments to reduce carbon omissions as they agreed to in the Kyoto Protocol, it is likely that they will turn to promoting reforestation in tropical countries, like Uganda, as a means of offsetting their carbon debt. This is already seen in activities of the FACE Foundation in Kibale and Mt. Elgon National Parks. Uganda can lead African countries in efforts of this nature. We say this because the Ugandan Government has made remarkable steps towards economic growth, while at the same time developing multifaceted environmental protection schemes. Thus, the economic growth that has occurred in Uganda over the last decade has been balanced, at least in part, by measures to protect the environment. These measures are formalized in the Constitution of Uganda (1995), the National Environment Statute (1995), the Wildlife Statute (1996), and the creation of the National Environment Management Authority (NEMA), to mention just a few. Furthermore, Protected Areas currently account for 14% of Uganda's total land area (Howard *et al.*, 2000). As a result, Uganda has the potential to become an African leader in biodiversity conservation and environmental policy.

Second, understanding restoration processes is vital for Kibale given its history. As early as 1971, illegal destruction and encroachment occurred in the southern half of Kibale. In 1976, some 30 eviction orders were issued, but were never carried out. In 1983, the government again ordered settlers out of these encroached areas, and by 1984, it was estimated that 60% of the forest plots and 30% of the grassland plots had been abandoned. However, the situation soon reverted to the prior state and encroachment increased. On April 1, 1992, the government ordered settlers off the land and resettled all encroachers. Estimates of the number of people residing in the southern corridor vary dramatically. Based on aerial surveys counting houses, van Orsdol (1986) estimated that 8800 people were living in the southern corridor. A national census carried out in 1980 indicated that as many as 17 000 people were residing in Kibale. Baranga (1991) estimated 40 000 people, MISR (1989) reported some 60 000 people, and after the resettlement the National Environmental Management Authority (NEMA, 1997) estimated that 30 000 households, or approximately 170 000 people, were residing in Kibale. Whichever estimate one chooses, it is evident that a large number of people were residing in the southern corridor, and as a result forest was cleared and degraded. Understanding processes of forest restoration will aid in the management of this area.

To provide the Uganda Wildlife Authority with an understanding of how they could facilitate forest regeneration to promote a forest that would support high red colobus abundance, we conducted a 4-year study to evaluate regeneration of indigenous trees 10 years after the pine plantations had been harvested. Further, since initial regeneration rates appeared slow, we conducted an enrichment planting experiment in an area of the harvested plantation and quantified the value of this planting program to enhance regeneration.

Pine plantations were established in Kibale between 1953 and 1977 on areas that were previously forested lands. These lands had been cultivated by agriculturalists, but abandoned when rinderpest devastated the livestock populations in the area in the early 1900s. Once the plantation matured, native tree species and shrubs invaded the understory and these were not removed by the plantation managers. With Kibale becoming National Park, management plans changed, the plantations were harvested and the areas were left to regenerate to native forest. The Kanyawara plantation was harvested between March 1995 and April 1996 and the area was left to regenerate.

Two 200 × 100 m (2 ha) plots were established in the harvested pine plantation and each was divided into 50 20 × 20 m sub-plots. In one plot (experimental), 200 seedlings of four species (*Albizia grandibracteata*, *Celtis africana*, *Celtis durandii*, and *Milletia dura*, 50 of each species) were planted. One of each of these species was planted in each 20 × 20 m sub-plot. These four species were selected because the seedlings were readily available, they often colonize disturbed areas, and these species have high protein to fiber ratio in the young leaves. The remaining plot was left as a control; however, a similar system of sub-plots was established to aid in quantifying regeneration and to ensure that each area received similar treatment by researchers. For each of the seedlings, saplings, and mature tree species in these plots, we determined the species identity and measured height, Diameter at Breast Height (DBH; if above 1.3 m), and Diameter at Ground Height (DGH).

A total of 44 tree species were found in the two areas, with 42 species in the experimental plot, and 35 in the control plot, and there was a great deal of overlap (80%) in species composition. The species that were not found in both plots were overall rare (maximum density 2 stems/ha). All tree species that were used as enrichment plant species in the experimental plots were encountered in higher numbers among the populations of the naturally regenerating trees in both plots. After 4 years naturally regenerating stems were much taller ($\bar{x} = 8.47$ m, $n = 1597$) than the planted seedlings ($\bar{x} = 0.5$ m, $n = 400$). The range of heights of planted seedlings (0.43–1.72 m) was much lower than the height of the naturally regenerating trees (0.2–15 m), and many of these seedlings are now under the canopy

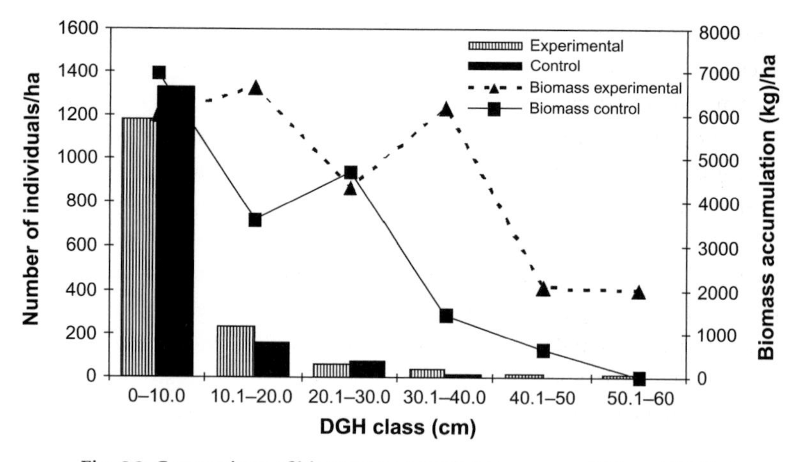

Fig. 6.2. Comparison of biomass accumulation among different DGH classes in experimental and control plots.

of trees that naturally regenerated. Counter to what was expected, the above ground biomass in the size classes of planted seedlings was greater in the control plot than in the experimental plot (Fig. 6.2).

Since the areas of Kibale that were grasslands in the 1960s and 1970s that were not converted to pine plantations are still largely fire-maintained grassland today, this study demonstrates that the use of fast-growing pine plantation species has facilitated the establishment and growth of indigenous tree species. However, this study also illustrates that, under these conditions, enrichment planting is not necessary for reforestation to progress at a reasonable rate.

The tree species that are regenerating naturally in these former pine plantations have leaves with higher protein to fiber ratios than found in the undisturbed forest. It is too soon to determine if the red colobus populations will increase with access to these regenerating areas containing these trees with high protein to fiber ratios; however, groups are frequently using these areas. We are monitoring the birth rates of groups that have access to these areas and groups that do not and we will soon have an experimental test of the prediction that access to foods high in protein and low in fiber will facilitate the growth of the endangered red colobus population.

DISCUSSION

The process outlined here could be coined *"Community Restoration"*, which we define as understanding determinants of animal abundance and plant community structure to such an extent that it is possible to modify the successional pathways of a plant community to arrive at a desired plant community structure for a set of animal species. We found no evidence in our study that there was a need for extra planting in our regenerating area; our goal of arriving at a desired plant community structure had been facilitated by the natural regeneration under pine plantations. In other cases, where regeneration is occurring from other forms of land conversion, supplementary planting may be a very effective tool, since we were able to see high survivorship in our seedlings. Supplementary planting may also be beneficial when attempting to manage for other types of animals (e.g., frugivores).

In Kibale it might be possible immediately to apply these notions to chimpanzees, since Balcomb *et al.* (2000) showed that chimpanzee density is related to the abundance of trees that produce large fruit and much is known about chimpanzee nutrition (Conklin-Brittain *et al.*, 1998). It is also possible that these concepts could be used in the opposite fashion, namely to push animals away from areas. For example, if tree species that were not eaten by elephants were planted in an area, once those trees matured, the area might be less attractive to herds than other areas (much is known about the nutritional requirement of Kibale's elephants) (Rode *et al.*, 2006). Given the growth in Kibale's elephant population, the park's increasingly isolated nature, and the damage that elephants do to the crops of the local communities (Naughton-Treves, 1998), this might be a long-term means of pushing elephants away from sensitive areas.

We expect that community restoration will be important for many tropical parks because many parks simultaneously contain areas of degraded forest and are attempting to protect endangered species that have limited geographical distributions. For Kibale, these questions are important because of the fire-maintained grasslands in the northern section of the park, degraded areas in the south, and endangered species of particular concern, such as the red colobus and chimpanzees. With regards to the grasslands in the northern sectors of Kibale, evidence suggests that fire frequency is decreasing in these areas (J. Lwanga personal communication) and thus, this habitat is changing. In this case, it should be recognized that taking no active management is a management decision in itself – just as significant to the landscape and the animal populations it supports as managing for trees that promote the population of a particular endangered species. We say this because, if Kibale's grasslands are not managed, many locally endangered grassland-dependent species will decline.

An important aspect to remember is that the ability of tropical forest managers and conservation biologists to use community restoration as a management tool requires access to long-term data. Researchers need to monitor changes in animal populations and habitat over decades, not simply the 2–3 year duration of a major grant or the 1 year of a typical graduate research project. Managers need to realize that getting funding for long-term monitoring is extremely difficult; much more difficult than to get funding for a series of studies to test particular academic hypotheses. As a result, a great deal of the long-term data that will be useful to managers will only arise if the managers are patient with researchers who often appear to be only addressing a narrow academic question and encourage researchers to integrate long-term aspects or monitoring protocols into their programs.

Our research on red colobus and restoration started almost two decades ago. However, our impressions of Kibale were greatly influenced by publications and the stories of the researchers who preceded us, such as Tom Struhsaker, Gil Basuta, John Kasenene, and Jerry Lwanga. With the aid of Tom Struhsaker, Lauren and Colin Chapman will soon have a 40-year record of climate change, plant phenology cycles, plant nutritional changes, red colobus behavior, and primate population density, and a 15-year data set on limnological patterns and regeneration. This illustrates that there is not only the need to encourage individuals to conduct long-term research, but there is a need to facilitate successive generations to keep working in specific areas. Such multi-generational research is facilitated greatly by the establishment and maintenance of long-term field sites.

SUMMARY

Human modification of ecosystems is threatening biodiversity on a global scale; effects are particularly severe in tropical rain forests where species diversity of many taxa is extraordinarily high. Closed-canopy tropical forest once covered 20% of Uganda's land area, but deforestation has reduced this to just 3%. In this chapter, we reported on a series of studies in Kibale National Park, Uganda to determine the critical factors that influence the abundance of endangered red colobus (*Procolobus rufomitratus*), and how we can facilitate forest regeneration in such a way to support high red colobus abundance. We found that the abundance of red colobus could be predicted by the protein to fiber ratio of available leaves, suggesting that managers could improve habitat quality for colobus by increasing the abundance of these foods. To explore the application of this idea, we conducted a 4-year study to evaluate regeneration of indigenous trees after a pine plantation had been harvested. Further, since initial regeneration rates appeared slow, we conducted enrichment planting using species with high protein to fiber leaves in one area. Tree biomass in the regenerating area was substantial in comparison to the grasslands on which the plantation was established originally, indicating that the use of fast-growing pine plantation species has facilitated the establishment and growth of indigenous tree species. However, enrichment planting did not promote regeneration and was not necessary for reforestation to progress at a reasonable rate. While the tree species that are regenerating in these former pine plantations have leaves with high protein to fiber ratios, it is too soon to determine if the red colobus populations with access to these regenerating areas will increase as would be predicted by the model predicting colobus biomass from the availability of leaves with high protein to fiber ratios. Our research emphasizes that, for tropical forest managers to be able to use information on the determinants of animal abundance and plant community structure as a management tool, long-term data are required.

ACKNOWLEDGMENTS

Funding for this research was provided by Canada Research Chairs Program, Wildlife Conservation Society, Natural Science and Engineering Research Council of Canada, and National Science Foundation. Permission to conduct the research discussed here was given by the Ugandan National Council for Science and Technology, and the Uganda Wildlife Authority. We thank Elizabeth Ross and Richard Wrangham for their efforts in pulling together the workshop that led to this chapter.

REFERENCES

Balcomb, S. R., Chapman, C. A., and Wrangham, R. W. (2000). Relationship between chimpanzee (*Pan troglodytes*) density and large, fleshy-fruit tree density: conservation implications. *American Journal of Primatology*, **51**, 197–203.

Baranga, J. (1991). Kibale Forest Game Corridor: man or wildlife? In *Nature Conservation: The Role of Corridors*., ed. D.A. Saunders and R. J. Hobbs. London: Surrey Beatty and Sons, pp. 371–375.

Chapman, C. A. and Chapman, L. J. (1999). Implications of small scale variation in ecological conditions for the diet and density of red colobus monkeys. *Primates*, **40**, 215–231.

Chapman, C. A. and Peres, C. A. (2001). Primate conservation in the new millennium: the role of scientists. *Evolutionary Anthropology*, **10**, 16–33.

Chapman, C. A., Chapman, L. J., Naughton-Treves, L., Lawes, M. J., and McDowell, L. R. (2004). Predicting folivorous primate abundance: validation of a nutritional model. *American Journal of Primatology*, **62**, 55–69.

Conklin-Brittain, N. L., Wrangham, R. W., and Hunt, K. (1998). Dietary response of chimpanzees and cercopithecines to seasonal variation in fruit abundance. II Macronutrients. *International Journal of Primatology*, **19**, 971–997.

FAO (2005). Global Forest Resources Assessment 2005: progress towards sustainable forest management. FAO Forestry Paper 147, Rome.

Ganzhorn, J. U. (2002). Distribution of a folivorous lemur in relation to seasonally varying food resources: integrating quantitative and qualitative aspects of food characteristics. *Oecologia*, **131**, 427–435.

Howard, P. C., Davenport, T. R. B., Kigenyi, F. W. *et al.* (2000). Protected area planning in the tropics: Uganda's national system of forest nature reserves. *Conservation Biology*, **14**, 858–875.

Milton, K., (1979). Factors influencing leaf choice by howler monkeys: a test of some hypotheses of food selection by generalist herbivores. *American Naturalist*, **114**, 363–378.

MISR (1989). *Settlement in Forest Reserves, Game reserves, and National Parks*. Kampala: Makerere University Press.

Naughton-Treves, L. (1998). Predicting patterns of crop damage by wildlife around Kibale National Park, Uganda. *Conservation Biology*, **12**, 156–168.

Naughton-Treves, L. and Chapman, C. A. (2002). Fuelwood resources and forest regeneration on fallow land in Uganda. *Journal of Sustainable Forestry*, **14**, 19–32.

NEMA (1997). *Kabarole District Environment Profile*. Kampala: NEMA.

Oates, J. F., Whitesides, G. H., Davies, A. G. *et al.* (1990). Determinants of variation in tropical forest primate biomass: new evidence from West Africa. *Ecology*, **71**, 328–343.

Paul, J. R., Randle, A. M., Chapman, C. A., and Chapman, L. J. (2004). Arrested succession in logging gaps: is tree seedling growth and survival limiting? *African Journal of Ecology*, **42**, 245–251.

Pomeroy, D. and Tushabe, H. (2004). *The State of Uganda's Biodiversity 2004*. Kampala: Makerere University Press.

Rode, K. D., Chiyo, P. I., Chapman, C. A., and McDowell, L. R. (2006). Nutritional ecology of elephants in Kibale National Park, Uganda, and its relationship with crop-raiding behaviour. *Journal of Tropical Ecology*, **22**, 441–449.

Struhsaker, T. T. (1975). *The Red Colobus Monkey*. Chicago: University of Chicago Press.

Struhsaker, T. T. (1997). *Ecology of an African Rain Forest: Logging in Kibale and the Conflict between Conservation and Exploitation*. Gainesville: University of Florida Press.

van Orsdol, K. G. (1986). Agricultural encroachment in Uganda's Kibale Forest. *Oryx*, **20**, 115–117.

JEREMIAH S. LWANGA AND G. ISABIRYE-BASUTA

7

Long-term perspectives on forest conservation: lessons from research in Kibale National Park

The manner in which humans use tropical rainforests has far-reaching consequences for the diversity of the world's terrestrial species, because tropical rainforests support more species than any other terrestrial ecosystem. Unfortunately, tropical forests all over the world are threatened by human activities, including forest fragmentation and isolation, collection of non-timber forest products, poaching, fires, forest degradation, and deforestation (mainly through selective logging for timber). Such activities have the potential to disrupt the integrity and functioning of forest habitats, which in turn may lead to loss of species and some of the resources that tropical forests provide. This chapter reviews the impact of human activities in Kibale National Park, Uganda.

Most of the conservation research that we report on here was short term (1 to 2 years). It was also based on a narrow range of ecological variables, such as the response of animal species to logging. As a result, the research findings are sometimes contradictory and difficult to interpret (Chapman *et al.*, 2005).

Kibale National Park has been the center of both short- and long-term research on various aspects of forest ecology (Struhsaker, 1997). Human activities that have led to modifications of the Kibale ecosystems have been going on for a long time but the best documented are those related to selective logging from 1954 to about 1978 and the loss of large herbivores (mainly elephant, *Loxodonta africana* and buffalo, *Syncerus caffer*). Since the 1980s, other human-induced changes have taken place. These include loss of forest fragments adjacent to the main forest block, colonization of grasslands by forest, removal of exotic plantations in the park and reforestation in abandoned agricultural land. While these changes were occurring, the human population in areas adjacent to the park was increasing rapidly. This led to increasing adverse human–wildlife conflicts including crop-raiding and injury to humans by wild animals. We examine the impact of some of these activities on the Kibale ecosystems and their long-term consequences for tropical forest conservation.

EFFECTS OF SELECTIVE LOGGING

Long-term data on the potential influence of selective logging on various vertebrate populations have been collected in Kibale Forest since the 1970s. Taxa studied intensively include primates, birds, small mammals, elephants, and other ungulates. These have been used as case studies of the impact of human activities on forest ecosystems. Logging occurred from the 1950s to the 1970s, mostly in the northwest of the former Forest Reserve, and covered an area of roughly 74 km^2 of forest. Logging intensity varied considerably. For example, Compartment 13 (K13) was heavily logged (17 m^3/ha) while K14 was lightly logged (14 m^3/ha). Further trees and saplings were damaged in the felling and loading process.

Response of primates to selective logging and forest fragmentation

Primates are good indicators of forest health. Detailed studies of primate populations in various habitats in Kibale impacted by humans have been conducted for over 30 years. Initially, declines in primate populations, except for black-and-white colobus (*Colobus guereza*), were attributed to selective logging. However, recent studies have shown that population declines of red colobus (*Procolobus rufomitratus*) and blue monkeys (*Cercopithecus mitis*) also occur in old growth forests (Mitani *et al.*, 2000; Chapman *et al.*, 2000). Further studies (Watts and Mitani, 2002; Teelen, 2005) have shown that predation by chimpanzees is the major cause of the decline of red colobus population at Ngogo. Furthermore, studies from Budongo Forest have shown that selective logging is not always associated with declines in primate populations (Reynolds, 2005). It appears that primate population responses to forest habitat alterations depend on a number of factors, including history of the forest, tree species removed and the diet of the primate species. The results of these studies clearly indicate the need for long-term multidisciplinary studies in evaluating potential impacts of habitat alterations on the conservation of primate populations. For example, a massive tree die-back has affected the diet of red colobus monkeys

(Struhsaker, 1975; Struhsaker *et al.*, 1989). Similarly, forest regeneration in formerly grassland areas has created new sources of food (for example leaves of *Prunus africana*) for folivorous primates (Chapman *et al.*, Chapter 6).

A result of these various effects is that it is no longer as easy to attribute declines in primate densities to logging as has sometimes been claimed. Long-term research in undisturbed areas is needed to generate data that allow managers to make interventions informed by natural trends.

Response of elephants to selective logging

Selective logging is associated with change in ground vegetation cover. In Kibale, selective logging stimulated the growth of dense ground cover dominated by *Acanthus* and other herbaceous plants. Studies have indicated that elephants frequent areas dominated by *Acanthus*. In so doing, elephants suppress tree regeneration through trampling and selective destructive feeding on tree seedlings, saplings, and poles (Struhsaker, 1997). For example, forest areas that were logged heavily over 30 years ago in Kibale are experiencing suppressed regeneration compared with similarly logged areas in Budongo Forest, which currently have no elephants and are regenerating well (Reynolds, 2006). The implication is that, in areas with elephants, forest managers should avoid creating large gaps while logging. Furthermore, discussions of the effects of selective logging need to take into account the numbers and types of herbivores within each forest type (Struhsaker, 1997; Reynolds, 2006).

Response of birds to selective logging

As with primates, comparisons of bird communities in logged and unlogged forests have sometimes been made on the assumption that the forests were similar to each other before logging took place. But, as Plumptre *et al.* (2001) pointed out, this assumption is highly unlikely in a spatially heterogeneous forest. Studies of the responses of bird communities to selective logging must therefore be interpreted with care.

Overall, in Budongo and Kibale a slight increase in the number of bird species has been found in logged compared to unlogged forest, but the dynamics of this comparison varied with forest type, size of logged patch, and bird feeding guilds among other factors. For example, in Budongo there was a high overlap (72%–82%) in bird species between logged and unlogged forests. By contrast, in Kibale, only relatively intact patches within the logged forest supported similar species to those in the unlogged forest, while heavily logged patches supported fewer bird species. Species diversity did not necessarily represent conservation success, because in Kibale, bird species classified as forest specialists, particularly crevice and hole nesters, had poor or no breeding in logged forest, even though they were present. The same was found in Budongo for Nahan's francolin (*Francolinus nahani*; Sande, 2001).

In terms of feeding guilds, species that eat fruits or seeds were barely affected by logging either in Budongo or Kibale, but in Kibale large frugivorous birds appear to require relatively large intact patches in forest from which fruiting trees have been removed. Furthermore, sallying insectivores appear to have been particularly vulnerable to logging in Kibale, although not in Budongo. In conclusion, many forest birds can survive in selectively logged forests and only a few species that depend on primary forest appear to suffer from selective logging (Plumptre *et al.*, 2001).

Difficulties in interpreting the impact of logging arise from several sources. Most importantly, it is hard to control for variables other than logging, such as tree species composition, aspect of slope, location within the forest, management history and altitude, as well as the intensity of logging itself. Second, the ecological niches of many bird species are still not well known. Third, forests change over time for reasons unconnected with logging, as illustrated by changes in tree species composition in the unlogged forest of K30 sector in Kibale Forest over the last 30 years (Struhsaker *et al.*, 1989). Future studies may have to deal with single species and/or feeding guilds.

Response of small mammals to selective logging

The impact of selective logging on the small mammal communities in Kibale forest was summarized by Struhsaker (1997). These studies reveal complexities that demand considerable knowledge of natural forest ecology in order to interpret the effects of logging. At least 26 small rodent species were recorded in forested habitats of Kibale. More species were found in the logged compartments than in unlogged compartments of mature forest. Apparently, this was due to the invasion of logged compartments by rodent species that are typical of colonizing bush and thickets and were probably responding to the opening up of the forest canopy as a result of logging. The relative densities of small rodents also varied between logged and unlogged forest, generally being higher in the logged forest. Furthermore, rodent densities were generally more stable in unlogged than in logged forest.

The factors responsible for the differences are poorly known, but food availability and rainfall appear to be influential. For example, in 1993 rodents were more abundant in the unlogged mature forest than in the logged forest (trap success 63% vs. 47%, respectively). This was apparently a result of two tree species in unlogged forest that produced an exceptionally large seed crop that was fed on by the small rodents. With regards to unlogged forest, rodents were positively correlated with ground vegetation cover. Rodent studies in Kibale thus indicate that other ecological factors, in addition to selective logging, interacted to influence both species richness and population densities.

Response of other ungulates to selective logging

Kibale is inhabited by at least nine ungulate species apart from elephants (Struhsaker, 1997). Nonetheless, only three, red and blue duikers (*Cephalophus harveyi* and *C. monticola*), and bush pigs (*Potamochoerus porcus*), have been studied in relation to logging (Struhsaker, 1997). Low densities render the rest difficult to study. A negative correlation exists between duiker abundance and the distance of a logged forest compartment from the mature unlogged forest, suggesting that the latter acts as a source of duikers dispersing into logged compartments. This observation suggests that, during logging operations, it is important to preserve intact areas of forests.

In general, duiker abundance was at least twice as high in the unlogged forest as in any of the logged compartments, strongly suggesting that logging has a negative impact on duiker abundance. The noise generated during logging was enough to drive duikers away from the vicinity of the logging operation (Struhsaker, 1997). There was no obvious relationship between logging intensity and duiker abundance, probably due to confounding factors such as poaching (Struhsaker, 1997) and habitat variation prior to logging. However, the interpretation of results obtained using different field study methods poses a serious constraint. For example, using duiker dung counts to estimate duiker abundance proved problematic. In some logged compartments duiker abundance obtained from direct observations was not correlated with the abundance of dung piles. In the logged forest, where duikers were encountered very infrequently using direct census methods, dung counts gave higher densities than those in the unlogged forest, where duikers were known to be more abundant (Struhsaker, 1997). A likely explanation is that duikers in logged forest compartments use cut trails more often to avoid the dense ground vegetation cover associated with logged forest. On the other hand, duikers in the unlogged forest compartments might not use cut trails as often because the ground vegetation cover there is sparse. This implies that dung piles are more likely to be missed in unlogged than in the logged forest. These results are complicated further by the fact that hunting pressure was known to be lower for the unlogged forest (near the field station) than for most of the logged compartment (located further away from the field station).

Using pig diggings to assess the impact of logging on the pig population, Nummelin (1990) did not detect differences in pig abundance between logged and unlogged forest compartments. However, this does not mean that logging was unimportant, since pig densities were probably affected by other factors also. First was slope: pigs rarely forage on upper and middle slopes where the ground is hard most of the time, and these were more common in unlogged areas. Second, a particularly important food-plant for pigs, *Palisota* spp., is most abundant in the valley bottoms. Third, poaching of bush pigs by members of the local communities has never been quantified; its impact on pig populations in various habitats is therefore unknown.

Studies of the responses of pigs and duikers to selective logging are thus difficult to interpret because logged and unlogged areas vary in several dimensions other than logging, such as the slope or nature of the food supply. There may be additional complicating factors that have not yet been considered, such as the presence of large buttressed trees (for protection and nursing of young) and the type of predators present in different habitats.

Influence of selective logging on forest structure and composition

The most obvious impact of selective logging on forest structure is the creation of large gaps. Selective logging also leads to reduction in the density of large stems, since in the logging process mainly large trees are removed.

Natural forest regeneration occurs in tree gaps created by tree falls. Gaps constitute microclimates necessary for the survival of seedlings and saplings of most forest tree species. Factors that influence regeneration in forest gaps are summarized by Struhsaker (1997). Research in Kibale indicates that gaps in the selectively logged forest are larger and more numerous and more uniform in age than those in unlogged forest. What are the implications of these differences for forest regeneration in Kibale National Park?

In heavily logged compartments, regeneration is slow or arrested. High browsing pressure and trampling by elephants have been suggested to be partly responsible. For example, elephants frequented gaps in logged forest more than would be expected based on proportional representation by area. Again, elephants were found to have damaged 25% of all saplings and poles along elephant paths. Small rodents, which are more abundant in logged than in unlogged forest, have also been implicated in the slowing down of regeneration, since many of them are seed and seedling predators. Creation of large gaps through selective logging alters the structure of the forest canopy, leading to the growth of dense floor cover and the effects noted above, such as elephants slowing regeneration by maintaining a dense herb layer. These findings emphasize that forests differ in many variables other than merely whether they have been logged.

EFFECTS OF FOREST FRAGMENTATION ON PLANT AND ANIMAL COMMUNITIES

Forest fragmentation occurs when continuous forests are divided into smaller patches of varying sizes and degrees of isolation. In the case of Kibale Forest, the once extensive forest has been fragmented on both the eastern and western boundaries leaving behind many small forest patches (Hamilton, 1984; Chapman *et al.*, 2002). As forest patches become smaller, we run the risk of losing some animal and plant species. Research from Kibale and nearby forests clearly shows that, even within the same species, the responses to fragmentation differ. Factors that are believed to determine persistence or extinction in forest fragments, such as dietary flexibility vs. specialization, or fragment size and distance between fragments, do not seem to explain all the variation.

For example, mangabeys (*Lophocebus albigena*) do not live in any of the forest fragments around Kibale National Park, but they are common in forest fragments further east in Uganda in the Lake Victoria basin. Similarly, blue monkeys in Kibale do not occupy forest fragments, yet they are present in some of the forest fragments around Budongo Forest. Chapman and Peres (2001) suggested that the survival of primates in fragments might be determined by the nature of the habitat that surrounds the forest fragment. These factors remain unknown, but need to be investigated if fragmented primate populations can be conserved as meta-populations (see Goldberg *et al.*, Chapter 8). In the case of plants, Chapman and Onderdonk (1998) reported that, in forest fragments where primate fauna has been reduced greatly, the density and richness of seedings were lower than in the intact forest. Such studies point to a complex set of interactions between forest size, animal seed dispersing agents, and the plant communities. Human activities that affect one will have effects on the others.

Forest fragmentation and risk of diseases to primates

Increased contact between humans and non-human primates as a result of forest fragmentation may elevate the risk of disease transmission between the two groups. At Kibale, increased forest patch degradation and the presence of humans strongly influenced the prevalence of parasitic gastrointestinal nematodes in red colobus monkeys (Gillespie and Chapman, 2006). Similarly, Struhsaker (2002) reports the presence of *Giardia* and *Entamoeba*, similar to those in humans, in three omnivorous cercopithecine monkeys of Kibale. Although the direction of pathogen transmission is not clear in these studies, both strongly suggest the possibility of transmission of pathogens from humans to non-human primates and vice versa. The risk of disease transmission between humans and great apes is particularly high, since apes are phylogenetically close to humans, and is liable to be promoted by forest fragmentation (Gillespie *et al.*, 2004).

Forest fragmentation and isolation

Forest fragmentation occurs when continuous forests are divided into smaller patches, and the intervening matrix of lands dominated by human activities creates barriers to movements of animals and plant propagules (Howe and Miriti, 2004). In principle, tropical rainforests are self-sustaining, i.e., they have the capacity to maintain themselves without human intervention (Bruenig, 1996). Nonetheless, this is only true if forests are large enough, in which case the best way to conserve them is to do nothing and let nature take its course. Unfortunately, this strategy is no longer possible for many forests because they exist mostly as isolated patches. In East Africa, even forests such as Kibale covering 776 km^2 are not large enough for the effective conservation of animal species with large home range requirements such as elephants and leopards, *Panthera pardus* (Struhsaker, 2002).

FOREST SUCCESSION IN KIBALE GRASSLANDS

About 15% of Kibale National Park was grassland in the 1970s (Wing and Buss, 1970). Presently, many of these grasslands are being colonized by tropical forest via three main pathways (active planting: Struhsaker, 2003; regeneration in former exotic plantations: Chapman and Chapman, 1996, Fimble and Fimble, 1996, Kasenene, 2007; and natural succession without human intervention: Lwanga, 2003).

Replanting a tropical rainforest can be effective (Matsuzawa and Kourouma, Chapter 17), but it tends to be costly and challenging. For example, Struhsaker (2005) cites a case where about $5 million were spent to plant and maintain an area of 32 km^2 for 9 years, and the forest that resulted from this effort was still species poor.

Regeneration in former exotic plantations in Kibale can provide nurse crops for many indigenous forest tree species (Chapman and Chapman, 1996; Fimbel and Fimbel, 1996; Kasenene, 2007). Kasenene's (2007) study shows that pine and cypress plantations provided better conditions than eucalyptus for the establishment of indigenous forest tree species. Therefore, conifer plantations (which in Kibale were established on grasslands) can be used to facilitate reforestation of degraded lands.

However, such elaborate interventions are not always necessary. At Ngogo, Kibale National Park, Lwanga (2003) demonstrated that simple protection of grasslands from fire led to accelerated forest succession. Thus, given the scarcity of funds for conservation, the most cost-effective and perhaps most efficient way of establishing tropical forests may be to allow natural succession through the prevention of fire.

Little research has been conducted on the effects of colonizing forests on animal communities. At Ngogo, duiker abundance in a colonizing forest that was not greatly affected by poachers was higher than in the old growth near the research camp, which was also relatively safe from poachers (Lwanga, 2006a). This observation is consistent with Wilkie and Finn's (1990) finding that, in the Ituri Forest (Democratic Republic of the Congo), secondary forests replacing slash-and-burn cultivation produce a higher density of hunted animal species than old growth forest. Regarding primates, Lwanga (2006b) sighted all eight diurnal primate species of Kibale in both colonizing and old growth forests, but black and white colobus monkeys were more abundant in colonizing forest than in the old growth forest while the converse was true for chimpanzees. These observations underscore the need not to write off colonizing forests. Colonizing forests have an important role to play, especially in the conservation of animal species that are adapted to disturbance. As they mature further, they also become important for animal species that are adapted to the old growth forest.

CROP-RAIDING

Crop-raiding and isolation are two important consequences of forest fragmentation. By isolation, we mean the introduction of artificial vegetation, such as farmland, adjacent to forest fragments. Forest-dwelling species such as elephants, bush pigs, chimpanzees, and monkeys can include food crops in their diets and can cause great losses to farmers adjacent to forests. Crop-raiding by primates can be controlled by guarding fields because primates are mainly active during the day, but the cost of guarding crops can exceed the value of the crops. Unfortunately, elephants and pigs crop-raid mainly during the night and are difficult to control by individual farmers. A study conducted at Kibale (Chiyo *et al.*, 2005) revealed that elephant invasions of farmland peak during the ripening of maize. In response to people's plight, park management constructed trenches to prevent elephants and pigs from entering agricultural land. However, as there was no evaluation of other potential interventions, the choice of trenches was not based on informed research. In our view, digging trenches may not have been the most cost-effective control mechanism in the long run. While this kind of intervention may be effective in protecting people's crops, it also isolates the forest further and the long-term effects of such isolation need to be investigated. Confining elephants in the forest may aggravate browsing pressure on the forest vegetation as the elephant population increases. These artificial barriers may also serve as traps for animals such as duikers, pigs, snakes, and frogs that do not have the ability to climb out of trenches. As an alternative to trenches, we encourage the management authorities to promote the growing of buffer crops such as tea (and other tree crops which are not palatable to crop-raiders) by local communities living adjacent to Protected Areas.

In spite of the current knowledge and management interventions, crop-raiding is still one of the most challenging problems facing researchers and wildlife managers. This is partly because interventions have paid relatively little attention to arboreal species such as primates and birds, yet in some localities these do the most damage. Additionally, both researchers and managers have tended to ignore crop-raiders originating outside Protected Areas, i.e., species living in forest patches, including primates, birds, and rodents. There are thus many opportunities for future research.

SUMMARY

Kibale has provided extensive opportunities for understanding factors promoting forest conservation. Numerous effects of selective logging and forest fragmentation have been found, and many are more complex than simple models would predict. This may be partly because, although the effects of logging have been documented over more than 30 years, much of the research on which this chapter is based comes from short-term studies (1 to 2 years). Long-term research is needed to clarify some of the complexities. In addition, there is still a substantial contribution waiting to be made by integrating long-term biological research with social science studies, for the benefit of wildlife managers.

REFERENCES

Bruenig, E. F. (1996). *Conservation and Management of Tropical Rainforests: An Integrated Approach to Sustainability*. Cambridge: Cambridge University Press.

Chapman, C. A. and Chapman, L. J. (1996). Exotic tree plantations and the rehabilitation of natural forests in Kibale Forest National Park, Uganda. *Biological Conservation*, **76**, 253–257.

Chapman, C. A. and Chapman, L. J. (1997). Forest regeneration in logged and unlogged forests in Kibale National Park, Uganda. *Biotropica*, **29**, 396–412.

Chapman, C. A. and Onderdonk, D. A. (1998). Forests without primates: primate/plant co-dependency. *American Journal of Primatology*, **45**, 127–141.

Chapman, C. A. and Peres, C. A. (2001). Primate conservation in the new millennium: the role of scientists. *Evolutionary Anthropology*, **10**, 16–33.

Chapman, C. A., Balcomb, S. R., Gillespie, T. R., Skorupa, J. P., and Struhsaker, T. T. (2000). Long-term effects of logging on African primate communities: a 28-year comparison from Kibale National Park, Uganda. *Conservation Biology*, **14**, 207–217.

Chapman, C. A., Chapman, L. J., Bjorndal, K. A., and Onkerdonk, D. A. (2002). Application of protein to fiber ratios to predict colobine abundance on different spatial scales. *International Journal of Primatology*, **23**, 283–310.

Chapman, C. A., Struhsaker, T. T., and Lambert, J. E. (2005). Thirty years of research in Kibale National Park, Uganda, reveals a complex picture for conservation. *International Journal of Primatology*, **26**, 539–555.

Chiyo, P. I., Chochrane, P. E., Naughton, L., and Basuta, G. I. (2005). Temporal patterns of crop-raiding by elephants: a response to changes in forage quality or crop availability? *African Journal of Primatology*, **43**, 48–55.

Fimble, R. A. and Fimble, C. C. (1996). The role of exotic conifer plantations in rehabilitation of degraded tropical forest lands: a case study from the Kibale Forest in Uganda. *Forest Ecology and Management*, **81**, 215–226.

Gillespie, T. R. and Chapman, C. A. (2006). Prediction of parasite infection: dynamics in primate metapopulations based on attributes of forest fragmentation. *Conservation Biology*, **20**, 441–448.

Gillespie, T. R., Greiner, E. C., and Chapman, A. C. (2004). Gastrointestinal parasites of the guenons of Western Uganda. *Journal of Parasitology*, **90**, 1356–1360.

Hamilton, A. C. (1984). *Deforestation in Uganda*. Nairobi, Kenya: Oxford University Press.

Howe, H. F. and Miriti, M. N. (2004). When seed dispersal matters. *Bioscience*, **54**, 651–660.

Kasenene, J. M. (2007). Impact of exotic plantations and harvesting methods on the regeneration of indigenous tree species in Kibale Forest, Uganda. *African Journal of Ecology*, **45**, 41–47.

Lwanga, J. S. (2003). Forest succession in Kibale National Park, Uganda: implications for forest restoration and management. *African Journal of Ecology*, **41**, 9–22.

Lwanga, J. S. (2006a). The influence of forest variation and possible effects of poaching on duiker abundance at Ngogo, Kibale National Park, Uganda. *African Journal of Ecology*, **44**, 209–218.
Lwanga, J. S. (2006b). Spatial distribution of primates in a mosaic of colonizing and old growth forest at Ngogo, Kibale National Park, Uganda. *Primates*, **47**, 230–238.
Mitani, J. C., Struhsaker, T. T., and Lwanga, J. S. (2000). Primate community dynamics in old growth forest over 23.5 years at Ngogo, Kibale National Park, Uganda: implications for conservation and census methods. *International Journal of Primatology*, **21**, 269–286.
Nummelin, M. (1990). Relative habitat use of duiker, bush pigs, and elephants in virgin and selectively logged areas of Kibale Forest, Uganda. *Tropical Zoology*, **3**, 111–120.
Plumptre, A. J., Dranzoa, C., and Owiunji, I. (2001). Bird communities in logged and unlogged African forests. Lessons from Uganda and beyond. In *The Cutting Edge: Conserving Wildlife in Logged Tropical Forests*, ed. R. A. Fimbel, A. Grajal, and J. G. Robinson. New York: Columbia University Press, pp. 13–238.
Reynolds, V. (2005). *The Chimpanzees of Budongo Forest: Ecology, Behaviour and Conservation*. Budongo Forest Reserve. In *Primates of Western Uganda*, Oxford, UK: Oxford University Press.
Reynolds, V. (2006). Threats to, and protection of, the chimpanzees of the Budongo Forest Reserve. In *Primates of Western Uganda*, ed. N. E. Newton-Fisher, H. Notman, J. D. Paterson, and V. Reynolds. New York: Springer, pp. 391–403.
Sande, E. (2001). The ecology of the Nahan's Francolin *Francolinus nahani* in Budongo Forest Reserve, Uganda. Ph.D. thesis, Makerere University, Uganda.
Struhsaker, T. T. (1975). *The Red Colobus Monkey*. Chicago: Chicago University Press.
Struhsaker, T. T. (1997). *Ecology of an African Rainforest*. Gainesville, Florida: Florida University Press.
Struhsaker, T. T. (2002). Strategies for conserving national parks in Africa with a case study from Uganda. In *Making Parks Work: Strategies for Conserving Tropical Nature*, ed. J. Terborgh, C. van Schaik, L. Davenport, and M. Rao. Washington: Island Press, pp. 97–110.
Struhsaker, T. T. (2003). Evaluation of UWA-FACE Natural High Forest Rehabilitation Project in Kibale National Park, Uganda. A report prepared for the center for applied biodiversity science of Conservation International and for the FACE Foundation.
Struhsaker, T. T. (2005). Conservation of red colobus and their habitats. *International Journal of Primatology*, **26**, 525–538.
Struhsaker, T. T., Kasenene, J. M., Gaither, J. C. Jr, Lassen, N., Musango, S., and Bancroft, R. (1989). Tree mortality in the Kibale forest, Uganda: a case study of dieback in a tropical rainforest adjacent to exotic conifer plantations. *Forest Ecology and Management*, **29**,165–185.
Teelen, S. (2005). The impact of hunting by chimpanzees *(Pan troglodytes)* on demography and behavior of Red Colobus Monkeys (*Procolobus rufomitratus*) at Ngogo, Kibale National Park, Uganda. Ph.D. dissertation,Yale University, New Haven.
Watts, D. P. and Mitani, J. C. (2002). Hunting behavior of chimpanzees at Ngogo, Kibale National Park, Uganda. *International Journal of Primatology*, **23**, 1–28.
Wilkie, D. S. and Finn, J. T. (1990). Slash-and-burn cultivation and mammal abundance in the Ituri forest, Zaire. *Biotropica*, **22**, 90–99.
Wing, L. D. and Buss, I. O. (1970). Elephants and forest. *Wildlife Monographs*, **19**.

TONY L. GOLDBERG, THOMAS R. GILLESPIE, AND
INNOCENT B. RWEGO

8

Health and disease in the people, primates, and domestic animals of Kibale National Park: implications for conservation

In the not-too-distant past, infectious disease was viewed as akin to fire, earthquake, and tornado in its propensity to impact wild primates. Outbreaks were considered inherently unpredictable, "sweeping through" primate populations, wreaking havoc, and then subsiding. Primates were generally thought to rebound, such that the overall effect was a "blip on the radar," a transient reduction in population numbers.

The last approximately 10 years have demonstrated the "disease as natural disaster" paradigm to be woefully inaccurate. Infectious disease has emerged as a major threat to primate conservation. The case of Ebola virus and its devastating effects on chimpanzees (*Pan troglodytes*) and gorillas (*Gorilla gorilla*) in Gabon and Congo is perhaps the most dramatic example, with some estimates of local population declines above 80% (Leroy *et al.*, 2004; Bermejo *et al.*, 2006). Other pathogens such as *Bacillus anthracis* (the causative agent of anthrax), polio virus, and yellow fever virus have also caused epidemic mortality in apes and monkeys, to the extent that they are now seen as important drivers of primate population declines (Chapman *et al.*, 2005; Leendertz *et al.*, 2006; Nunn and Altizer, 2006).

Despite these dramatic examples, the majority of primate pathogens probably exert chronic, sublethal effects on primates in the wild (most parasitic protozoa, helminths, and arthropods probably fall into this category). Although researchers are paying increasing attention to such agents, most studies to date have been either "prevalence surveys" or comparisons of prevalence across locations or habitat types. It remains unclear to what extent endemic pathogens and the chronic diseases they cause might regulate primate populations, impact primate demographics, and alter primate behavior (Chapman *et al.*, 2005).

Emerging infections threaten global human health as much as they do primate conservation. Novel infectious diseases are emerging today in human populations at an accelerated rate worldwide, and the trend shows no signs of abating. Microbes thought to be on the brink of extinction decades ago remain tenaciously endemic, both because of gaps in surveillance and because the pathogens themselves have shown a surprising ability to evolve. Pathogens such as HIV, West Nile virus, SARS coronavirus, and influenza virus emerge and re-emerge with disquieting regularity, in some cases causing epidemic or pandemic mortality. Globalization, climate change, and increased contact with reservoir species through agricultural intensification and natural resource exploitation all drive this trend (Daszak *et al.*, 2000; Daszak *et al.*, 2001; Woolhouse and Gowtage-Sequeria, 2005).

Although humans have always shared habitats with non-human primates, the dynamics of human–primate interactions have changed dramatically in the recent past. Within the last several decades, humans have altered primate habitats irrevocably, exploiting tropical forest ecosystems at an ever-increasing rate as the material and economic needs of expanding human populations grow (Cowlishaw and Dunbar, 2000). Many primates today live in habitat mosaics of farmland, human settlements, and forest fragments, and in isolated Protected Areas such as National Parks (Marsh, 2003; Fig. 8.1). Human influences in the form of roads, hunting, and climate change are reaching even into the last remaining strongholds of primate biodiversity in such countries as the Democratic Republic of the Congo, Brazil, and Indonesia (Chapman and Peres, 2001).

Infectious disease emergence is an unfortunate and unanticipated consequence of these ecological changes. Indeed, a full 75% of emerging human infectious diseases are zoonotic or have recent zoonotic origins, with wildlife, livestock, and domestic carnivores serving as sources of infection (Taylor *et al.*, 2001). Comparative epidemiological analyses indicate that an ability to cross any species barriers enhances the probability that a pathogen will be classified as "emerging" (Cleaveland *et al.*, 2001; Taylor *et al.*, 2001; Woolhouse and Gowtage-Sequeria, 2005). This realization, combined with a sense of urgency about anthropogenic environmental change, has spawned a series of new disciplines bearing such names as "conservation medicine" or "ecosystem health," complete with dedicated societies, journals, and international meetings (Daszak *et al.*, 2004).

Fig. 8.1. Black-and-white colobus in Rurama forest fragment, approximately 1 km from the western edge of Kibale National Park. Primates in such locations must run a gauntlet of threats each day, from aggressive dogs to habitats scattered with pathogens.

The "Kibale ecoHealth Project" represents an attempt to bring this nascent and evolving paradigm to bear on infectious diseases shared among primates, people, and domestic animals in the region of Kibale National Park, Uganda. Founded in 2004, the project takes a "place-based," epidemiological approach to understanding the interrelationships among primate health, human health, and the health of domestic animals in an anthropogenically altered environment (Fig. 8.2).

Kibale is an ideal location for studying infectious disease and its relationship to primate conservation. The Kibale EcoHealth Project builds directly on long-term research that has taken place in Kibale over the last several decades. The Kibale EcoHealth Project works closely with the Kibale Monkey Project, for example, benefiting from groundwork and ecological

Fig. 8.2. The Kibale ecoHealth Project logo, meant to represent the ecological interdependency of human, primate, and domestic animal health.

data collected over the past approximately 15 years on primate populations inside and outside of the park. The Kibale EcoHealth Project collaborates with the Kibale Chimpanzee Project to identify risks for disease transmission between chimpanzees and people inside and outside of the park. The Kibale EcoHealth Project works extensively with local communities around Kibale – an arrangement that would not have been possible without the positive community relations that have been built over the years by efforts such as the Kasisi Project. Although the Kibale EcoHealth Project itself is relatively young, it continues in the tradition of long-term research and successful conservation that has helped make Kibale one of the premier tropical forest locations in the world for research, conservation, education, and sustainable development.

Kibale is also well suited to the goals of the Kibale EcoHealth Project because of the varied types and degrees of disturbance that characterize locations inside and outside of the park. Kibale's habitats range from essentially undisturbed core forest to highly disturbed and unprotected forest fragments outside of the park proper. This gradient of anthropogenic disturbance facilitates a "natural experiment" approach to studying primate disease. The Kibale region is also unfortunately representative of locations throughout the tropics, where conservation efforts and land-use change intersect. The approach of the Kibale EcoHealth Project is therefore both basic and "translational:" it attempts to elucidate fundamental processes of disease ecology at the same time that it identifies practical intervention strategies.

The process by which pathogens cross species barriers and eventually cause persistent health problems involves a complicated series of steps, each with its own (usually low) probability (Wolfe *et al.*, 2007). For example, diseases that find their way into new species do not always possess the ability to spread among individuals within that new species, and diseases that can spread within a new species sometimes fail to perpetuate. Nevertheless, the initial "jump" from one species to another is the critical step, because interrupting the process of transmission between species eliminates the possibility of any "downstream effects." For this reason, the Kibale EcoHealth Project focuses its scientific efforts on understanding how anthropogenic factors lead to increased pathogen transmission between species.

Previous research in the Kibale area has demonstrated that certain types of anthropogenic disturbance alter the prevalence of gastrointestinal helminths in wild primates. For example, Gillespie *et al.* (2005) documented an increased prevalence and richness of gastrointestinal helminth and protozoan infections in red-tailed guenons (*Cercopithecus ascanius*) in logged forest compared to undisturbed forest. Gillespie and Chapman (2006) investigated a series of forest fragments and showed that the density of tree stumps, an "honest indicator" of human encroachment, was a strong predictor of the prevalence of gastrointestinal helminths in red colobus (*Procolobus rufomitratus*). Chapman *et al.* (2006) further demonstrated that red colobus in forest fragments near Kibale suffer increased gastrointestinal parasitism with helminths as a result of nutritional stress, and that the effects of stress and parasitism can lead to population declines. Salzer *et al.* (2007) demonstrated that red colobus in forest fragments, but not in undisturbed forest locations, are infected with *Cryptosporidium* and *Giardia*, two gastrointestinal protozoa known to be important for human and livestock health.

Documenting increased prevalence of parasites in anthropogenically disturbed habitats is important for demonstrating the negative effects of anthropogenic processes such as logging and forest fragmentation on primate health. However, it is already well known that fragmentation and other related alterations to forest ecosystems threaten primate populations; documenting yet another negative impact of such processes does little to guide intervention. For this reason, the Kibale ecoHealth Project focuses its efforts on elucidating the ecological mechanisms underlying increased disease risk to primates in disturbed habitats.

Since 2004, the Kibale EcoHealth Project has targeted a series of forest fragments near the western edge of Kibale National Park (approximately 1 km from the park boundary and approximately 1 to 2 square kilometers in size). Behavioral observations of primates in these fragments have yielded intriguing (although often anecdotal) information that may help explain why primates in anthropogenically disturbed forests are at risk of exchanging pathogens with people and domestic animals.

Primates in the forest fragments near Kibale appear to be especially aggressive. In 2005, we documented an unusual case of aggression by red colobus monkeys against a raptor (Goldberg *et al.*, 2006). During a polyspecific association of red colobus, black-and-white colobus (*Colobus guereza*), and red-tailed guenons in Rurama forest fragment, a large raptor flew overhead, eliciting alarm calls and general consternation. Although the raptor left without incident, a pearl-spotted owlet (*Glaucidium perlatum*) coincidentally (and unfortunately) flew into the vicinity 2 minutes later. In their "hyper-aggressive state," the largest male red colobus in the group pounced on the bird and killed it (but did not eat it). This behavior had not been observed in red colobus in the undisturbed sections of Kibale, despite nearly 10 000 hours of observation.

The raptor-directed aggression documented during this incident may reflect a more general phenomenon. Since 2004, Kibale EcoHealth Project field assistants have been threatened or attacked on several occasions by red colobus males in forest fragments. Interviews with local farmers indicate that aggressive interactions with red colobus males occur regularly in the forest fragments near Kibale. On one memorable occasion, a large male red colobus in Kiko forest fragment descended from a tree and ran in a very determined manner across approximately 50 meters of open pasture to attack a field assistant who was observing the group with binoculars (defensive maneuvers with an aluminum clipboard and a hand-held GPS unit averted injury to either party).

The frequent aggression displayed by red colobus males in forest fragments is not surprising, considering the hostile interactions that primates routinely have with people and domestic animals in such habitats. Kibale EcoHealth Project field assistants frequently have observed children throwing rocks and sticks at monkeys. Active guarding of crops against crop raiding by primates is a common activity in villages near forest fragments, and interviews with local people suggest that parents sometimes keep their children home from school to engage in this activity. Dogs routinely are chained in fields near forest fragment edges to protect crops against marauding primates.

A particularly intriguing incident occurred in 2006, when Kibale EcoHealth Project field assistants observed a domestic dog attacking and wounding a juvenile red colobus monkey. The attack occurred when the red colobus group came to the ground to cross an open space between patches of trees. The dog caught the monkey in its jaws and drew blood. If not for the intervention of the observers, the monkey would most likely have been killed (the dog released its prey when the observers intervened, and the monkey was seen back with its group soon thereafter; its long-term fate is unknown). Several incidents involving "rogue male" chimpanzees have been reported in the Kibale area in which aggressive male chimps have attacked and severely wounded human children.

The anecdotes described above suggest that direct contact between primates and people/domestic animals may occur more regularly than has previously been appreciated in the Kibale region. This realization has serious implications for disease transmission risk. In West Africa, where hunting is common and where people routinely contact primates and their bodily fluids, blood-borne viral pathogens are transmitted regularly from primates to people (Wolfe *et al.*, 2005). Systematic hunting of primates by people does not occur in the Kibale region, but other aggressive encounters between people and primates could have similar effects on infectious disease dynamics. Domestic animals could play critical roles in enhancing human–primate disease transmission. Dogs, for example, may serve as intermediate hosts for the transmission of blood-borne viruses between primates and people. The high prevalence of HIV in Uganda raises further concerns; AIDS renders a large proportion of the human population in the Kibale region immunocompromised and thus particularly susceptible to opportunistic infections such as novel zoonoses.

Aggressive interactions between primates and other species are leading to the decline and extinction of primates in the forest fragments near Kibale. In January 2006, black-and-white colobus in Kiko 1 forest fragment disappeared. Interviews with local farmers indicated that dogs had most likely killed the entire group of monkeys. Approximately 6 months later, the final four remaining red colobus in Kiko 1 fragment also disappeared. Again, interviews indicated that these monkeys were likely killed by packs of dogs. At the time that these primate species disappeared, rates of forest clearing within Kiko 1 fragment were inordinately high; virtually no forest remains in the location of the fragment today. The association of deforestation, interspecies aggression, and local primate extinction in this forest fragment is almost certainly not coincidental. It is, however, unfortunately representative of the state of affairs in the unprotected forests of the Kibale region, as well as in other locations throughout the tropics where human population expansion fuels the destruction of forests.

Primate–human conflict in the Kibale region is not always as overt or dramatic as the above examples might suggest. Much primate–human–domestic animal conflict plays out on larger spatial scales and over more protracted time frames. For example, primates must come to the ground to cross open spaces between suitable habitat patches within forest fragments. Such movement exposes them to domestic animal feces when they move through pastures, or to human feces when they move through forest fragment edge habitats, where people often defecate when working in fields. Primates may well become infected with human and domestic animal pathogens in this manner.

Crop-raiding by primates is another "risky" behavior that may also increase infectious disease transmission. To raid crops, primates must often cross pastures, dodge chained dogs and packs of roving dogs, and avoid being injured by farmers or their children who guard crops actively. In addition, people in the Kibale region have adopted cultural practices that may increase the disease-associated risks of crop-raiding. "Maize daubing" is a case in point. Farmers in the Kibale region apply a mixture of sand and cattle dung to ears of maize on the edges of fields bordering forest fragments. This practice is meant specifically to deter crop-raiding by red-tailed guenons, a species that may be an obligate crop-raider in forest fragments and that has a penchant for maize (Naughton-Treves *et al.*, 1998). The practice may, however, inadvertently facilitate the transmission of gastrointestinal pathogens from livestock to primates.

Although conflict between primates and humans in the Kibale region can have negative disease-related effects, so too, paradoxically, can activities related to conservation. Chimpanzees in Kibale's Kanyawara community have been studied continuously for 20 years, and chimpanzees in Kanyanchu community are a major focus of "ecotourism" in Uganda. Both research and tourism have contributed in overwhelmingly positive ways to the conservation of Kibale's chimpanzees, enhancing the long-term survival of the apes by increasing their scientific and economic value, respectively. Nevertheless, Goldberg *et al.* (2007) demonstrated that research and tourism may enhance transmission of the common gastrointestinal bacterium *Escherichia coli* between chimpanzees and the humans who work with them. Chimpanzees in Kibale tended to harbor bacteria that were more similar genetically to the bacteria of field assistants and ranger-guides who work with chimpanzees and spend long hours in their habitats than to bacteria of local villagers whose interactions with chimps are limited. Moreover, chimpanzees harbored *E. coli* resistant to multiple antibiotics used by people in the region, indicating that microbes or their genes can "diffuse" from humans to chimpanzees even in the best of conservation circumstances. "Friendly" activities such as research and tourism may therefore not be without their own disease-related risks.

The overall picture emerging from the Kibale EcoHealth Project is that a variety of specific human, primate, and domestic animal factors impact disease transmission risk among these species (Table 8.1). Importantly, results to date indicate that direct contact between species is not necessary for interspecific disease transmission. Indeed, most transmission of gastrointestinal pathogens between people and primates is probably indirect and environmental. Pathogens such as *Cryptosporidium*, *Giardia*, and *E. coli* readily contaminate water and soil and may persist in wet areas. Human, primate, and domestic animal contact with common environmental sources of infection may explain many of the trends that the Kibale EcoHealth Project has documented. Figure 8.3 presents

Table 8.1 *Generalized factors hypothesized to increase disease transmission risk among people, primates, and domestic animals in fragmented forests in western Uganda*

Human factors	Primate factors	Domestic animal factors
Direct agonistic interactions with primates, such as guarding crops against crop-raiding	Direct encounters with people and domestic animals as a result of home range overlap	Hunting of primates (dogs)
Indirect agonistic and deterrent interactions with primates (e.g., "maize daubing")	Crop-raiding and incursions into human settlements	Grazing at the edges of forest fragments, near fields separating primate habitat patches
Forest clearing, extractive forestry, and encroachment into primate habitats	Movement across landscapes frequented by livestock, especially on the ground	Contamination of physical environment with environmentally persistent pathogens
Utilizing water sources located within primate home ranges		

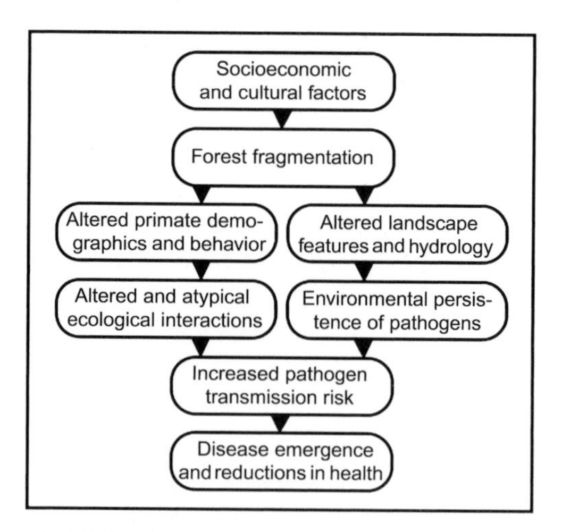

Fig. 8.3. Working conceptual model for how forest fragmentation may lead to infectious disease emergence in primates, people, and domestic animals in the Kibale region and other ecologically similar locations.

a working conceptual model for how forest fragmentation might ultimately impact infectious disease emergence and reductions in the health of humans and primates in the Kibale region and in ecologically similar areas.

If human behavior is indeed a strong force influencing the transmission of pathogens between primates and people, then targeted interventions should be possible. Making people aware of the disease-related risks of their activities, and providing alternatives, could go far towards reducing interspecific disease transmission and improving human health, animal health, and primate conservation. Examples such as "maize daubing" present obvious opportunities for intervention, but less apparent control points may also exist. For example, in the forest fragments near Kibale, restricting the movement of domestic dogs could increase primate survival and decrease pathogen transmission risk. Locating human latrines away from forest fragment edges and digging strategically placed wells could also improve human and primate health by reducing gastrointestinal disease transmission through the environment. In the case of chimpanzee research and tourism, burying human feces (if it must be buried) away from low-lying, wet areas where it could contaminate streams might similarly reduce gastrointestinal pathogen transmission risk.

The fate of primates in the unprotected areas near Kibale National Park is unclear, as is the fate of primates throughout the tropics living in fragmented and disturbed habitats where rates of contact with humans and domestic animals are high. Without intervention, we should expect primates in such settings to undergo rapid local extinction, and we should further expect such extinctions to be accompanied by "spikes" in infectious disease transmission. Only with a detailed ecological understanding of how human behavior alters the dynamics of disease transmission among primates, people, and domestic animals can we design rational intervention strategies that contribute efficiently and effectively to primate conservation.

SUMMARY

Infectious disease represents a serious and growing threat to primate health and conservation, especially in anthropogenically disturbed habitats where primates interact at high rates with people and livestock. Research in and near Kibale National Park, Uganda, has demonstrated that anthropogenic factors alter both the prevalence of pathogens in primates and rates of transmission of pathogens between primates and other species. Since 2004, the Kibale EcoHealth Project has focused on primates living in forest fragments outside the Protected Areas of the park. Primates in these fragments engage in frequent agonistic interactions with people and domestic animals, ranging from direct contact (e.g., hunting by packs of dogs) to habitat overlap (e.g., crop raiding by primates, encroachment into primate habitats by people). Human–primate conflict is leading to local extinctions of primates from forest fragments near Kibale and to simultaneous increases in infectious disease transmission among primates, people, and domestic animals. Targeted interventions focused on specific human practices have the greatest potential to reduce pathogen transmission among species, thereby safeguarding animal health, human health, and primate conservation in the Kibale region and other ecologically similar areas.

ACKNOWLEDGMENTS

We gratefully acknowledge the Uganda Wildlife Authority and the Uganda National Council for Science and Technology for granting us permission to conduct research in the Kibale region, and we thank Makerere University Biological Field Station for providing facilities and logistic support. We thank the William and Flora Hewlett Foundation, the Geraldine R. Dodge Foundation, the Wildlife Conservation Society, The University of Illinois Center for Zoonoses and Infectious Disease Research, the University of Illinois Earth and Society Initiative, and the

Morris Animal Foundation for providing financial support for the Kibale EcoHealth Project and its related activities.

REFERENCES

Bermejo, M., Rodríguez-Teijeiro, J. D., Illera, G., Barroso, A., Vilà, C., and Walsh, P. D. (2006). Ebola outbreak killed 5000 gorillas. *Science*, **314**, 1564.

Chapman, C. A. and Peres, C. (2001). Primate conservation in the new millennium: the role of scientists. *Evolutionary Anthropology*, **10**, 16–33.

Chapman, C. A., Gillespie, T. R., and Goldberg, T. L. (2005). Primates and the ecology of their infectious diseases: how will anthropogenic change affect host–parasite interactions? *Evolutionary Anthropology*, **14**, 134–144.

Chapman, C. A., Wasserman, M. D., Gillespie, T. R. *et al.* (2006). Do food availability, parasitism, and stress have synergistic effects on red colobus populations living in forest fragments? *American Journal of Physical Anthropology*, **131**, 525–534.

Cleaveland, S., Laurenson, M. K., and Taylor, L. H. (2001). Diseases of humans and their domestic mammals: pathogen characteristics, host range and the risk of emergence. *Philosophical Transactions of the Royal Society of London B, Biological Sciences*, **356**, 991–999.

Cowlishaw, G. and Dunbar, R. (2000). *Primate Conservation Biology*. Chicago: University of Chicago Press.

Daszak, P., Cunningham, A. A., and Hyatt, A. D. (2000). Emerging infectious diseases of wildlife – threats to biodiversity and human health. *Science*, **287**, 443–449.

Daszak, P., Cunningham, A. A., and Hyatt, A. D. (2001). Anthropogenic environmental change and the emergence of infectious diseases in wildlife. *Acta Tropica*, **78**, 103–116.

Daszak, P., Tabor, G. M., Kilpatrick, A. M., Epstein, J., and Plowright, R. (2004). Conservation medicine and a new agenda for emerging diseases. *Annals of the New York Academy of Sciences*, **1026**, 1–11.

Gillespie, T. R. and Chapman, C. A. (2006). Prediction of parasite infection dynamics in primate metapopulations based on attributes of forest fragmentation. *Conservation Biology*, **20**, 441–448.

Gillespie, T. R., Chapman, C. A., and Greiner, E. C. (2005). Effects of logging on gastrointestinal parasite infections and infection risk in African primates. *Journal of Applied Ecology*, **42**, 699–707.

Goldberg, T. L., Gillespie, T. R., Rwego, I. B., and Kaganzi, C. (2006). Killing of a pearl-spotted owlet (*Glaucidium perlatum*) by male red colobus monkeys (*Procolobus tephrosceles*) in a forest fragment near Kibale National Park, Uganda. *American Journal of Primatology*, **68**, 1–5.

Goldberg, T. L., Gillespie, T. R., Rwego, I. B., Wheeler, E. R., Estoff, E. E., and Chapman, C. A. (2007). Patterns of gastrointestinal bacterial exchange between chimpanzees and humans involved in research and tourism in western Uganda. *Biological Conservation*, **135**, 527–533.

Leendertz, F. H., Pauli, G., Maetz-Rensing, K. *et al.* (2006). Pathogens as drivers of population declines: the importance of systematic monitoring in great apes and other threatened mammals. *Biological Conservation*, **131**, 325–337.

Leroy, E. M., Rouquet, P., Formenty, P. *et al.* (2004). Multiple Ebola virus transmission events and rapid decline of central African wildlife. *Science*, **303**, 387–390.

Marsh, L. K., ed. (2003). *Primates in Fragments: Ecology and Conservation*. New York: Kluwer Academic/Plenum Publishers.

Naughton-Treves, L., Treves, A., Chapman, C., and Wrangham, R. (1998). Temporal patterns of crop-raiding by primates: linking food availability in croplands and adjacent forest. *Journal of Applied Ecology*, **35**, 596–606.

Nunn, C. L. and Altizer, S. (2006). *Infectious Diseases in Primates: Behavior, Ecology and Evolution*. Oxford: Oxford University Press.

Salzer, J. S., Rwego, I. B., Goldberg, T. L., Kuhlenschmidt, M. S., and Gillespie, T. R. (2007). *Giardia* sp. and *Cryptosporidium* sp. infections in primates in fragmented and undisturbed forest in western Uganda. *Journal of Parasitology*, **93**, 439–440.

Taylor, L. H., Latham, S. M., and Woolhouse, M. E. (2001). Risk factors for human disease emergence. *Philosophical Transactions of the Royal Society of London B, Biological Sciences*, **356**, 983–989.

Wolfe, N. D., Daszak, P., Kilpatrick, A. M., and Burke, D. S. (2005). Bushmeat hunting, deforestation, and predicting zoonotic emergence. *Emerging Infectious Diseases*, **11**, 1822–1827.

Wolfe, N. D., Dunavan, C. P., and Diamond, J. (2007). Origins of major human infectious diseases, *Nature*, **447**, 279–283.

Woolhouse, M. E. and Gowtage-Sequeria, S. (2005). Host range and emerging and reemerging pathogens. *Emerging Infectious Diseases*, **11**, 1842–1847.

ROSIE TREVELYAN AND CLIVE NUTTMAN

9

The importance of training national and international scientists for conservation research

Conservation of biological diversity depends upon a critical mass of dedicated, well-trained people. Developing countries such as Uganda require a cadre of nationals who can work at all levels from policy to science so that their respective country's conservation challenges can be met. Long-term research and training programs can play an important role in creating such a cadre as well as forging international links that bring in extra expertise and resources. Their success is best measured by how many nationals become research leaders and how many become effective mentors and trainers.

The International Union for the Conservation of Nature's World Conservation Strategy recognized in 1980 the need for training conservation leaders in tropical countries. This is reiterated in the Convention on Biological Diversity (1992), with Article 12 specifically highlighting the need for research training in developing countries. Nevertheless, nationals from developing countries still perceive a need for more training in a range of topics including conservation biology, monitoring and evaluation, fundraising, and project design (Bonine *et al.*, 2003).

Although the tropics is host to a large number of field stations (the World Register of Field Centres lists a good majority of them), much of the well-funded research and training that is carried out in the tropics is driven by international rather than by national researchers. Long-term programs can therefore play a role in assisting tropical field stations and their collaborators meet national training and research priorities and develop conservation capacity among national scientists. Field training provides a critical element in a biologist's education that cannot be achieved in the university classroom. This is true for both temperate and tropical countries, although it is commonly the case that there are fewer resources available in developing countries for them to run field courses. Research and training are complementary activities: training courses rely on research results, preferably with the researchers themselves explaining them, while new research projects are often inspired by project questions of inquisitive trainees. Long-term research programs additionally provide vital training for nationals through supporting MScs and PhDs.

Makerere University Biological Field Station (MUBFS) is one of Africa's successful examples of a field center that hosts long-term training programs from a variety of institutions. There are three main ingredients that contribute to the success of MUBFS as one of Uganda's foremost field centers. The first is the location itself: MUBFS sits within Kibale forest which offers a variety of habitats and associated flora and fauna, all of which are easily accessed by foot directly from the field station via the extensive trail system. The second is local expertise. Long-term research in Kibale has provided MUBFS with a wealth of knowledge and an active research community (and their field assistants) who make important contributions to field training directly or indirectly. The final ingredient is the facilities that MUBFS offers in the form of accommodation, fresh water, laboratory and teaching rooms, library, and mains or generator electricity. These result largely from USAID investments in the past and now need to be maintained through user fees from researchers and educational groups.

The Tropical Biology Association (TBA) is an educational NGO that is helping to build a critical mass of trained and motivated conservation biologists in Africa. It was established to expand the regional coverage of other training organizations focused largely on South and Central America such as the Smithsonian Institute (Wemmer *et al.*, 1993) and the Organization for Tropical Studies. The TBA and MUBFS began working together in 1994 and, since then, they have run 21 month-long field courses at the field station. The courses have provided hands-on training in tropical ecology and conservation coupled with project design using Kibale forest as the outdoor classroom. Because of its situation in the forest, MUBFS provides a 24-hour educational experience and makes the logistics of field classes easy. It also encourages students to plan their own time schedules when carrying out independent research projects, since they are not relying on other people for transport or on guides.

The TBA courses at MUBFS take the unique approach of training African biologists alongside biologists from Europe in equal numbers. Twenty-four students attend each course representing anything between

12 and 18 different countries. This cultural mix provides an excellent forum for sharing ideas between the two continents and raising awareness about different approaches to conservation. The courses are designed to plug an obvious gap that universities and other organizations tend to leave in terms of practical skills and approaches to problem solving and ecological interpretation. At the same time, the courses provide invaluable first-hand experience of the complexity of conservation issues where local community needs and the sustainable management of natural habitats are being addressed simultaneously. Case studies based around the work of local researchers and conservation programs comprise important components of the courses. The target group from Africa is conservation biologists who have done a first degree and wish to complete post-graduate training or who are working in institutions responsible for the management and monitoring of natural resources. Students from outside Africa tend to be recent graduates or Master's students who hope to use the course to decide upon their future career paths. The TBA invites international and local experts to teach on the courses, offering a rich array of topics and teaching approaches. Since 1994, the TBA courses have trained 68 Ugandan biologists, several of whom return regularly as teachers.

Running a continuing series of field courses at one site means that training organizations develop a detailed knowledge of the area and create a broad repertoire of projects. In addition, such programs have great potential to contribute to long-term monitoring and research being carried out at field centers. There are several monitoring opportunities that regular training programs could take up in Kibale Forest including the sites where the forest is regenerating, as well as other forest monitoring plots. In addition to building on existing knowledge, long-standing training programs can contribute new ideas and catalyze new research projects through inviting scientists from different disciplines than those the field center already covers. For example, during the last 15 years, the TBA has brought in 219 teachers from 133 institutions in 25 countries. This has expanded the network of international researchers that MUBFS can draw upon and has helped raise the profile of MUBFS globally as a research and training station. Several TBA teachers have returned to Kibale to conduct research programs of their own.

One reason that field courses are successful is that students are more likely to learn through active learning approaches in the field rather than through the traditional lecture-based education in the classroom (for a review of active learning see Michael, 2006). For example, TBA field courses include a combination of natural history workshops, discussion seminars, class field exercises, and supervised project work. Seminars encourage participants to share their experiences and provide a significant forum for both teachers and students from a broad array of backgrounds to learn from each other. Field exercises are a successful method of teaching ecological principles by involving the class in designing a field experiment and collecting quantitative data to address specific concepts being addressed. Participants not only learn teamwork, they also gain experience in how to ask ecological questions and how to analyze and present results. Field exercises demonstrate that there is rarely a simple result to ecological experiments and that interpretation of data requires careful consideration of an array of factors. TBA field courses often tackle ideas that have not been tested at the field stations before, and a number of field exercises, and indeed projects, end up being published in peer-reviewed journals.

The best way to gain expertise in field research is to do it oneself. Students who undertake projects on field courses learn how to frame research questions and choose appropriate methods as well as analyze and interpret results. On TBA courses for example, students work in pairs, following the rule of not working with someone from their own country, and each project team is given individual guidance at each step of the project cycle. Almost without exception, participants say that the projects are one of their most important learning experiences on their TBA course, providing them with key skills that they fully expect to use back home afterwards. For many, this is the first time they have designed their own research project, even if they are undertaking a Master's, and this helps build confidence in their own abilities as independent researchers. Participants also find that the statistics training using their own data sets is particularly useful. Projects additionally impart skills in scientific writing and presenting results orally: projects are written up as research papers and presented in a seminar at the end of the course. Abstracts of TBA projects are hosted on the TBA website and full project reports are sent to field stations, universities, and research institutions around Africa. Project reports can be a useful resource to institutions. They generate ideas for research projects and give examples of different experimental and analytical approaches to field research. Projects often tackle subjects that are not part of the mainstream research programs at field stations and hence can make useful contributions to the knowledge base of the area.

The involvement of nationals and the use of case studies are key components of tropical field courses and are important means of building understanding of local conservation and management issues among participants (see Hills *et al.*, 2006 for an example of this process working in coastal management). It is certainly the TBA's experience that conservation teaching has the biggest impact on trainees through site visits, case studies, and through discussions with conservation practitioners. Visits to the villages around the edges of Kibale forest during the "community day" on TBA courses provide first-hand experience of community-based approaches to conservation and human–wildlife conflict. Europeans are not taught such practical aspects of tropical conservation in their universities, while most Africans have much to contribute by offering details of the different approaches in their home countries. Uganda Wildlife Authority staff take an active part in TBA courses to give their perspective on managing National Parks (the courses also visit Queen Elizabeth National Park) and their work with local communities.

EXCHANGING TEACHING EXPERTISE

Field courses do not just improve the skills and knowledge of the students. Teachers attending courses learn new ideas and approaches from other course faculty and gain experience in field teaching. The exchange of expertise on TBA courses at MUBFS works across a multitude of levels. The international setting builds confidence and communication skills among teachers who have only taught at the national level. Local field assistants also benefit from the interaction with TBA teachers at the same time as imparting their knowledge to the course students and faculty. Twenty-four Ugandans have instructed on TBA courses and many have reported that they use ideas and materials gained during the course to teach students back at their home institutions.

INCREASING THE SKILLS OF AFRICAN SCIENTISTS IN PUBLISHING AND PROPOSAL WRITING

Long-term research programs can play an important role in imparting skills in scientific writing and mentoring national scientists through the publishing process. Nevertheless, there is still an imbalance between African authors and authors from the US or Europe in the international biological literature (Harrison, 2006). The TBA has, therefore, responded to requests from its African partners to develop practical workshops that teach skills in scientific writing, publishing, and proposal development. Between 2006 and 2007, 35 conservation scientists from 17 Ugandan institutions including universities, government departments, NGOs, and the Uganda Wildlife Authority have attended these specialist workshops.

All participants are given the TBA skills series "scientific writing" and "fund-raising guidelines" as well as Power Point presentations so that they can pass on the training to others.

Training workshops are a useful tool to identify specific barriers that prevent national scientists from publishing their research work. Feedback from TBA trainees is that the perceived barriers to publishing include lack of confidence, not knowing whether the investigator has collected enough data of the right quality, and lack of experience in the writing skills required for journals. Since the lack of mentors tends to be a major constraint identified by national scientists, those researchers who have long-established links in Kibale can play a very constructive part in mentoring Ugandan researchers and sharing the process of writing papers with them.

FOLLOW-UP SUPPORT: LIFE BEYOND THE FIELD STATION

The TBA has created a unique follow-up support program to assist course and workshop alumni to apply their new skills once back at their home institutions. The program is also a means by which others working in biodiversity research and conservation can recruit collaborators, students, or employees from a network of alumni and teachers who have relevant expertise. The TBA provides direct support to former trainees in the form of scholarships, grants, and subscriptions to conservation journals. Since beginning this program in 2002, TBA has raised over half a million dollars to sponsor MSc students to pursue graduate degrees at universities throughout Africa and Europe. Of more than 20 Master's students supported, many have completed their degrees and are now enrolled in PhD programs while others have taken up key jobs as conservation managers. The TBA also provides advice through its mentoring program to assist young conservation biologists to write papers or funding proposals.

The highest number of requests for follow-up support from TBA Ugandan alumni is in the form of scholarships for MSc or PhD studies (Fig. 9.1). Requests for jobs, project funding and fellowships follow in frequency. Monitoring this kind of information helps TBA match its efforts in providing relevant support and information to its past course trainees.

Much of the work of the TBA follow-up support program happens through its internet resource center. Hosted on TBA's website, an online directory of funding sources contains information and web addresses for over 185 grants, fellowships and training opportunities relevant to African conservation biologists. In the first 6 months of its launch it attracted over 600 users. The TBA bulletin board is an electronic jobs,

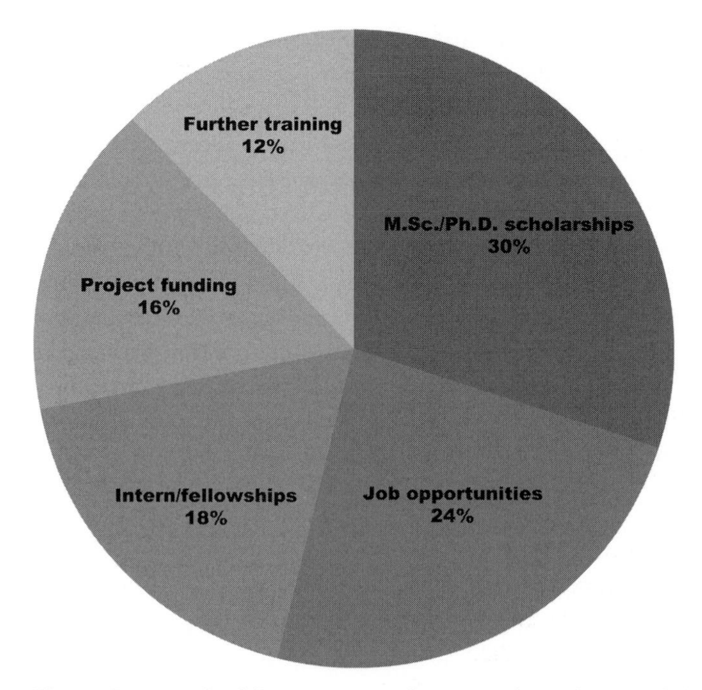

Fig. 9.1. Requests for follow-up support from Ugandans who attended TBA courses between 1994 and 2006.

news, and opportunities directory. Updated monthly, it contains recent job adverts, conferences, and resources being offered. The resource center additionally contains guidelines for preparing funding proposals and writing scientific papers and abstracts from TBA projects that can be downloaded by the user.

While benefiting greatly from the TBA's follow-up activities, a good number of course alumni have set up their own groups for in-country mutual support. National TBA groups exist in Kenya, Nigeria, Cameroon, Ghana, and Sudan with new groups currently forming in Malawi, Ethiopia, and Tanzania. The aims of the national TBA groups are to strengthen links and create a mutual support network between young conservation biologists as well as catalyze new conservation partnerships. Each group has its own web-forum, and the TBA bulletin board acts a hub for them all.

MEASURING IMPACT

The impacts of graduate-level training programs may not be seen until several years down the line and will probably be influenced by further experiences that the trainee gains over time. The TBA looks at both a short-term and long-term time frame for measuring its impact.

Short-term impact: number of trainees and quality of training

The short-term impact of TBA training activities is measured through the number of scientists trained and the quality of the training. In 13 years, the TBA has delivered 41 month-long field courses that have trained 484 African conservation biologists of 23 nationalities alongside 502 European and American biologists. One hundred and two Ugandan biologists have benefited from the field courses and specialist training workshops, and many work in key government and non-government conservation organizations in Uganda.

The TBA measures the trainees' perception of the quality of the training and how useful the outcomes of the training will be in their work back home through anonymous questionnaires using both quantitative and narrative responses. Questions cover a variety of measures from the quality and content of the teaching to the balance of the curriculum between fieldwork and lectures or seminars. The TBA courses consistently gain high scores for the time allocated between lectures and fieldwork and for the teaching quality on each. Participants also rate very highly the experience they gain of designing and executing a project, often commenting that this is where they learned the most during the course. This feedback reinforces the importance of "learning by doing" which is a key approach to field course teaching. Narrative feedback includes without exception how much participants gain from the multinational nature of the course, which cannot be found in an average university setting. People value the sense of belonging within the global conservation community and expect to continue their connections beyond the course.

Long-term impact

The second method the TBA uses to assess the impact of its training courses is to monitor where trainees go after the courses and how they use the skills or contacts gained during the training. Of the 484 Africans who attended TBA courses from 1994 to 2006, 94% are in regular contact with the TBA. This rate increases to 100% for those who attended the courses between 2002 and 2006. This exceptional figure allows the TBA to monitor the progress of its former trainees over time as well as to provide continuing assistance and mentoring to these individuals.

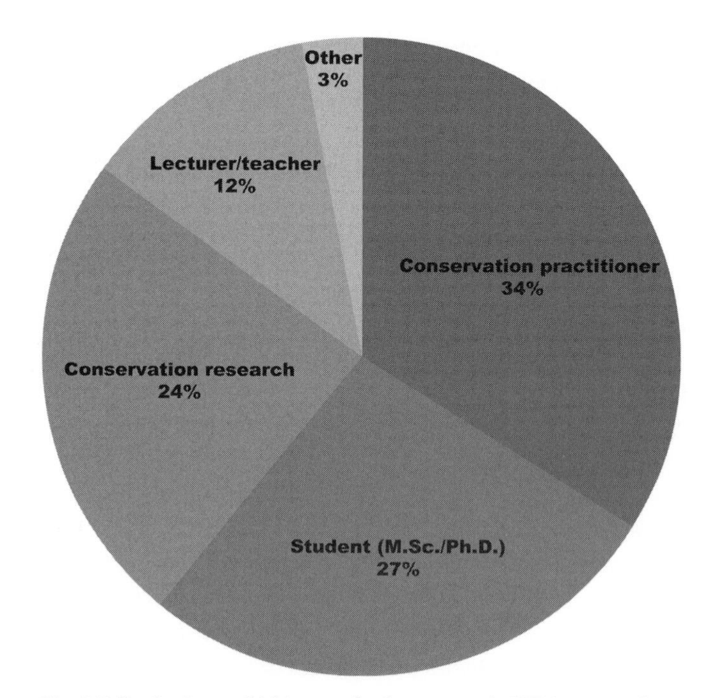

Fig. 9.2. Professions of Africans who have attended TBA courses between 1994 and 2006. Data obtained from a 94% contact rate.

An analysis of Africans who attended courses between 1994 and 2006 shows that 97% are engaged in conservation biology: 27% are studying for M.Sc.s or Ph.D.s, 12% are teaching, and the rest have a variety of jobs as conservation practitioners or researchers (Fig. 9.2). Looking specifically at Ugandans, the greatest number (24%) work in forest ecology after their courses (Fig. 9.3). Sixteen percent and 13%, respectively, work in conservation generally or community conservation specifically. Nineteen percent work in environmental education and 6% work in aquatic ecology.

The type of institutions that trainees join after their courses yields useful information on the degree of impact that they might be having on conservation in their home countries (Fig. 9.4). These data tell us whether people are working in fields such as policy, advocacy, or education and awareness and how directly they might be impacting conservation through their work. Several Ugandan alumni work in government departments (22% of alumni) such as the National Forest Authority, the Ministry of Tourism, Trade and Industry, and the Ministry of Education and hence are likely to be in positions to influence policy internally. A larger number of alumni (36%) work in national and international non-government

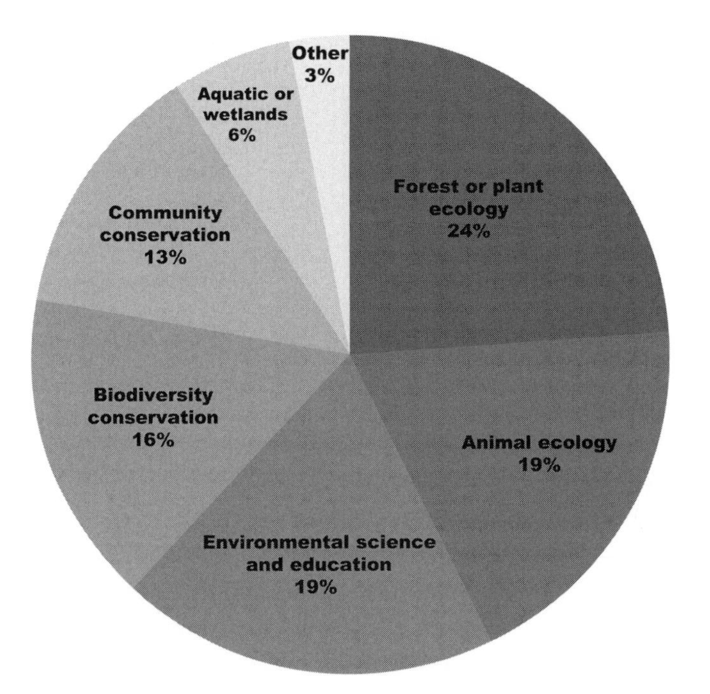

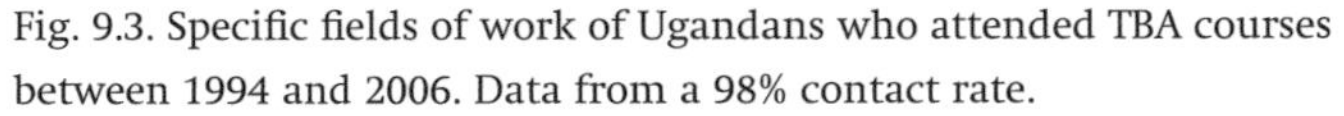
Fig. 9.3. Specific fields of work of Ugandans who attended TBA courses between 1994 and 2006. Data from a 98% contact rate.

organizations such as Nature Uganda, Wildlife Clubs of Uganda, the Institute of Tropical Forest Conservation, the Wildlife Conservation Society, and Care International. The same number work in universities, while 4% work in inter-government institutions such as the IUCN and International Council for Research in Agroforestry (World Agroforestry Center); 2% of Ugandan alumni work in business.

THE WAY FORWARD: HOW CAN LONG-TERM RESEARCH AND TRAINING CONTRIBUTE TO GREATER RESEARCH CAPACITY IN UGANDA?

The future of long-term research and conservation in Uganda requires a critical mass of national scientists who have the resources and institutional support they need to be effective. International projects can help develop this critical mass if they collaborate with Ugandan institutions in their priority research and training areas. Involving Ugandan researchers in responsible positions in projects provides an effective means of transferring expertise and building confidence. Crucially, this places them in a better position to provide training to others either through more

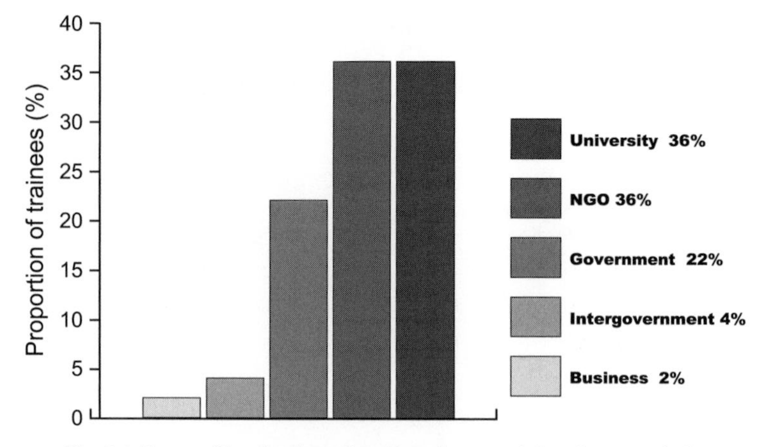

Fig. 9.4. Types of institutions that Ugandans work in who attended courses between 1994 and 2006. Data from a 98% contact rate.

university field courses or through mentoring the next generation of Ugandan researchers. It is becoming increasingly important for Ugandan scientists to have a bigger profile in the research community, and to do this they need to share their results through publications and other media. As well as increasing their profile, it is likely to attract more collaborators to their specific fields and expand their research outputs as a result. If all of this is achieved, long-term research programs in Uganda will have a better chance of being sustainable in the long run.

REFERENCES

Bonine, K., Reid, J., and Dalzen, R. (2003). Training and education for tropical conservation. *Conservation Biology*, **17**, 1209–1218.

Harrison, A. (2006). Who's who in conservation biology: an authorship analysis. *Conservation Biology*, **20**, 652–657.

Hills, J. M., Alcock, D., Higham, T. *et al.* (2006). Capacity building for integrated coastal management in Asia–Pacific: the case for case studies. *Coastal Management*, **34**, 323–337.

International Union for the Conservation of Nature, UN Environment Programme, and World Wildlife Federation (1980). *World Conservation Strategy: Living Resource Conservation for Sustainable Development*. Morges: IUCN.

Michael, J. (2006). Where's the evidence that active learning works? *Advances in Physiology Education*, **30**, 159–167.

Wemmer, C., Rudran, R., Dallmeier, F., and Wilson, D. E. (1993). Training developing-country nationals is the critical ingredient to conserving global biodiversity. *Bioscience*, **43**, 762–767.

World Register of Field Centres http://www.rgs.org/OurWork/Fieldwork+and+Expeditions/World+Register+of+FieldCentres accessed on September 5, 2007.

JOHN M. KASENENE AND ELIZABETH A. ROSS

10

Community benefits from long-term research programs: a case study from Kibale National Park, Uganda

The authors of this chapter have known each other for 20 years. For the past 10 years we have been partners in The Kasiisi Project, a project that invests in primary schools in the forest-edge communities around Kibale National Park. We represent the two sides of the Ugandan "aid coin:" expatriate donor (The Kasiisi Project) and Ugandan recipient (AFROKAPS). We share a commitment to common goals but, because of our different cultural backgrounds, we sometimes approach them in different ways. We recognize that our priorities may differ and we do not always agree with each other. Local pressures and community expectations on either side of the Atlantic shackle us both and misunderstandings occasionally arise, but 20 years have given us a foundation of familiarity and friendship that has enabled us to work effectively as a team. Together, we have built a successful project that now works with five schools and 3500 children.

This chapter looks at the ways in which long-term research programs in Kibale National Park (KNP) have brought benefits to local, national and international communities and how this, in turn, can have positive implications for conservation. Our main focus is on small community projects, started by mainly expatriate researchers and their families. We suggest that the biggest contribution of long-term research programs to these projects is the fostering of alliances between expatriates and nationals. We believe that our experience is probably typical of projects that involve western researchers interacting with tropical forest communities (Collins and Goodall, Chapter 14).

INTRODUCTION

The Kibale Forest Project (KFP) was started by Thomas Struhsaker in 1970 with a focus on the ecology and behavior of the red colobus monkey (*Procolobus rufomitratus*) (Struhsaker, 1975). Struhsaker selected the Kibale Forest Reserve for primate research because it harbored viable populations of these animals and their existence was guaranteed by the cultural respect the local people held for them. Other primate studies followed to investigate the ecology and behavior of blue monkeys (*Cercopithecus mitis*) (Rudran, 1978; Butynski, 1982), black-and-white colobus monkeys (*Colobus guereza Rueppel*) (Oates, 1974; Baranga, 1983) and chimpanzees (*Pan troglodytes*) (Basuta, 1987). In 1987 when the Kibale Forest Project was being wound up, it became imperative to establish a fully fledged research station to organize and direct the various research interests in the Kibale Forest Reserve. This resulted in the creation of the Makerere University Biological Field Station (MUBFS) in 1987. MUBFS coordinates a diversity of research programs studying various aspects of tropical forest ecology with an emphasis on long-term research (Fig. 10.1). In collaboration with the forest department authority and national and international universities, the original primate research has expanded its scope to include investigations of both the Kibale Forest ecosystem as a whole and the surrounding environment (Buhapa, 1994; Chapman and Chapman, 1996; Kasenene, 1984; Naughton-Treves, 1998; Van Orsdol, 1986). There have also been significant long-term and short-term sociological and ethno-botanical research projects outside the forest (Kakudidi *et al.*, 2000; Kasenene, 1998).

COMMUNITY BENEFITS

The communities that have benefited directly or indirectly from the research activities in KNP can be categorized as the international research and conservation community, the national Ugandan community, and the local community.

BENEFITS TO INTERNATIONAL RESEARCH COMMUNITY

Research stations like MUBFS, themselves a result of long-term research programs, can have implications for conservation beyond their own borders. The research opportunities in KNP and MUBFS have attracted international scientists and students. Researchers from America, Europe, the Far East, Australia, and other African countries have pursued advanced research programs and graduate degrees in KNP. MUBFS has

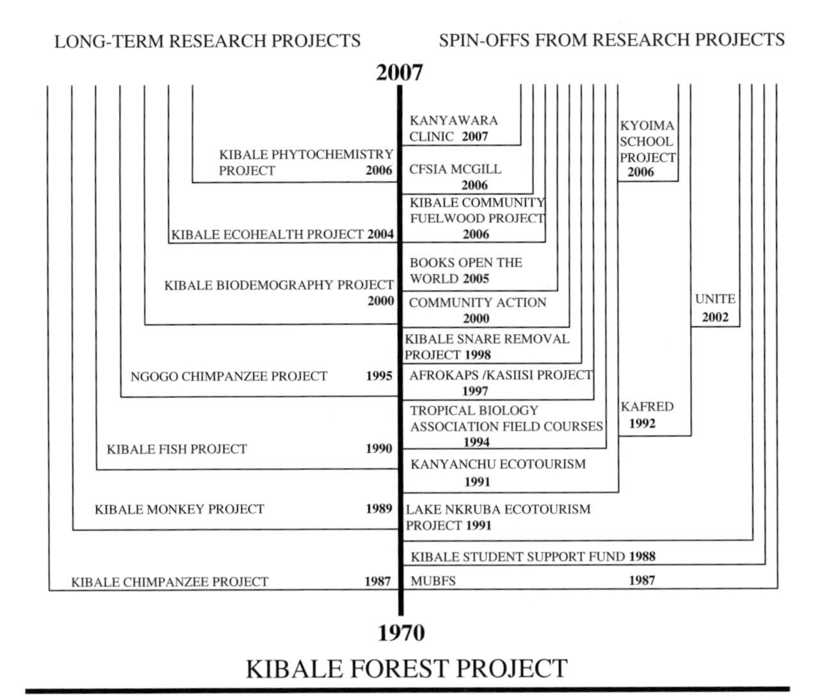

Fig. 10.1. A time line of current long-term research and conservation projects in Kibale National Park, research projects are shown on the left and conservation projects on the right.

proved an ideal place for the sharing of research experiences that lead to productive collaborations, particularly among those who went on to develop long-term research programs in the Park. Lessons learned in Kibale have informed conservation programs in other countries. Two former directors of MUBFS, Gary Tabor and Andy Grieser-Johns, went on to manage conservation programs in the USA and Vietnam. Other scientists from Kibale have developed research stations elsewhere in Uganda. The Institute of Tropical Forest Conservation in Bwindi Impenetrable Forest was founded by Tom Butynski, a primatologist trained in Kibale.

BENEFITS FOR THE NATIONAL/UGANDAN COMMUNITY

The international researchers who come to KNP, attracted by the Field Station, assist the cause of Ugandan conservation by supporting national scientists, the more so if they stay in Uganda to develop long-term studies. They often include Ugandan counterparts in their research programs and a number of Ugandans have received scholarships and placements in foreign universities (Trevelyan and Nuttman, Chapter 9). This type of support is very important for Ugandan scientists. It enables them to further their education and to benefit from international exposure. Foreign researchers in Kibale have contributed to the advancement of national science by teaching special subjects, related to their research topics, at Makerere University. They act as supervisors of postgraduate students and examiners of postgraduate courses, helping graduate students acquire new research knowledge and skills. The Makerere University, as an institution for teaching and research, has also benefited from the success of expatriate researchers in obtaining funds from big international donors: USAID, EU, WWF, and IUCN among others. In KNP this money has financed the research station at Kanyawara (MUBFS), The Kanyanchu Tourism and Training Center and the Ngogo Camp. All play important roles in conservation of the forest. Government departments and NGOs also use these research stations and training centers for seminars and workshops. Financial support for MUBFS comes from the user fees and rent charged to researchers, students, and organizers of seminars and workshops. The presence of long-term research programs, which attract new projects and increasing numbers of students, maximizes the income from these fees. Increasingly, the Uganda Wildlife Authority (UWA), the agency in charge of KNP, receives important research information, which it uses in developing conservation and management plans for the Park. Researchers also assist UWA's work by traversing large areas of KNP thus providing a patrol function, and their regular presence also contributes to control of illegal activities in the research zone.

BENEFITS TO THE LOCAL COMMUNITY

At the local level, communities benefit from the presence of MUBFS and the associated long-term research projects carried out in KNP in many ways. Figure 10.1 shows a selection of the current major community projects that have arisen as a direct result of long-term research programs. These can be divided into general economic benefits, e.g. employment, benefits that are directly tied to a specific research program, e.g. ecotourism at Kanyanchu, and initiatives that arise from the personal desire of researchers to do something about the compelling needs they see all around them, e.g. the Kasiisi Project.

General economic benefits

Research in the forest has contributed significantly to the local economy by providing employment. Local villages provide laborers for all construction work in the forest. The building of MUBFS, in particular, generated considerable income. In 1991 there were 170 people on the MUBFS payroll. These communities continue to provide housekeepers, gardeners, field research assistants, guides and interpreters, trail cutters, and collectors of plant specimens. A particularly important program employs, among others, former poachers, under careful supervision, to patrol the forest and remove illegal snares (Wrangham, Chapter 1). These jobs, a direct consequence of the presence of the field station and research being carried out in KNP, have a positive economic effect on the local community. In addition, encouraged by visiting researchers and supported by foreign grants, co-operatives producing handcrafts have started in the surrounding villages. Visiting field courses and the many visitors drawn by the long-term research programs provide a good market for crafts, which are also sold to raise money overseas for other community projects.

On a more informal level, researchers supply free transport to town, markets and hospitals. In times of medical emergency, they provide rapid transport to hospital. They also participate in community conservation programs including environmental education and sensitization. Examples include forest conservation and tree-planting programs, primary health care initiatives, soil and crop improvement, the use of fuel-efficient stoves and trials of non-traditional, vermin-resistant crops which give high yields in nutrient poor soil around KNP. The latter were selected to minimize damage from wildlife, thus helping to improve community relations with the National Park. Tree planting programs encourage the cultivation of fuel wood for domestic use and local markets, reducing pressure on natural recourses and providing income.

Benefits from specific research projects

Protected area managers sometimes complain that the field station generates too little of the applied research that they need for conservation. However, it is clear that pure research designed to answer academic questions can have a profound effect on the socioeconomic status of neighbouring communities while at the same time promoting the conservation of vital ecosystems. An example is the way that research by the Kibale Chimpanzee Project into the behavioral ecology of chimpanzees in Kanyawara led to the development of ecotourism at Kanyanchu (Mugisha, Chapter 11). This brought substantial economic benefits to local villages and conserved the Magombe wetlands. Trained field research assistants at Kanyanchu, learning about community conservation through their work, collaborated with relevant communities in the local town of Bigodi to save the nearby Magombe swamps. The result was the Kibale Association for Rural and Economic Development (KAFRED), a community-based organization that established a community ecotourism program based on the conservation and development of the Magombe Wetland. KAFRED provided many job opportunities for local people and worked to educate the relevant communities on the need to conserve vital natural resources in the Magombe Wetland. The revenue generated from Magombe ecotourism has helped the Bigodi community to build a secondary school, supply needy students from the Bigodi community with secondary school scholarships, support a maternity clinic at a Bigodi dispensary, and improve the condition of local primary schools. The women of Bigodi mobilized and formed groups to provide special services to the visitors including canteen services and non-traditional food and drinks. The men built restaurants to provide catering and accommodation services. In efforts to increase the value of the benefits from KNP and the wetland, the communities of Bigodi trained in the best methods of harvesting materials from the wild and in how to make quality crafts for sale to tourists.

Although the long-term research aims of the Kibale Chimpanzee Project did not seem to have an obvious connection to the economic status of the local people or to directly impact conservation of the Magombe swamps, it has clearly done both through its role in developing ecotourism at Kanyanchu. This shows that the boundary between basic/pure and applied research can be hazy. Both types of research should be encouraged as they offer useful information for redirecting and diversifying conservation options, while at the same time increasing the spectrum of benefits to relevant communities surrounding the Protected Areas.

Expatriate-funded community projects

There is a direct connection between long-term research and projects that benefit the communities bordering KNP, which can be seen in the programs begun by researchers and their families. The Kasiisi Project is this kind of project. At MUBFS, these aid projects are of two main types. The first is where a clear need is identified, rectified, and finished. For example, the family of research student Tara Harris paid for the flooring of two classrooms at Kanyawara Primary School. The second involves projects with longer-term, open-ended goals. Those to improve adult literacy and to encourage the use of fuel-efficient stoves are two examples (see below). The Kasiisi Project also falls into this second group.

Below, we describe four projects of the second type with emphasis on our own work with the Kasiisi Project (Table 10.1). We have seen how well

Table 10.1. *Examples of Community projects in Kibale National Park in 2007*

Community Project	Research Project	Description	Budget 2006–2007
The Kasiisi Project (1997) www.kasiisiproject.org	Kibale Chimpanzee Project (1987)	Educational support	$70 000
Books open the World (2005) www.booksopentheworld.org	Kibale Biodemography Project (2000)	Community libraries, adult literacy, women's groups	$11 536
The Kibale Student Support Fund (1988) www.kibalestudents.org	The Kibale Forest Project (1970)	Primary and secondary school scholarships	$9 000
The Kibale Community Fuel Wood Project (2006) www.chimp-n-sea.org	The Kibale Fish and Monkey Projects (1989), The Kibale Ecohealth Project (2004), The Kibale Chimpanzee Project (1987)	Cultivated fuel woods, fuel-efficient stoves	$39 203

these small projects can succeed. They get aid to the people who need it in a cost-effective and efficient way. They have proved easy to supervise because of their small size and, while they do not have the reach of larger NGOs, they can produce a better product for less money. An example can be seen at Kanyawara primary school where there are classrooms, funded by the Irish government and the Kasiisi Project, which were built at the same time. The government classrooms, which cost more (up to three times as much according to one source in the local education authority), are already crumbling, whereas the Kasiisi Project buildings are still in good repair.

We also discuss what we consider to be the most important factors for the success of these projects and the role that long-term research plays in maximizing their effect. Lists of donors to these projects can be accessed from their websites.

PROJECT DESCRIPTIONS

The Kasiisi Project

In 1997 children from Kanyawara, the village closest to MUBFS and the core research area, had no local school. They attended Kasiisi Primary School 8 km away (school 1, Fig. 10.2). In the same year Uganda introduced

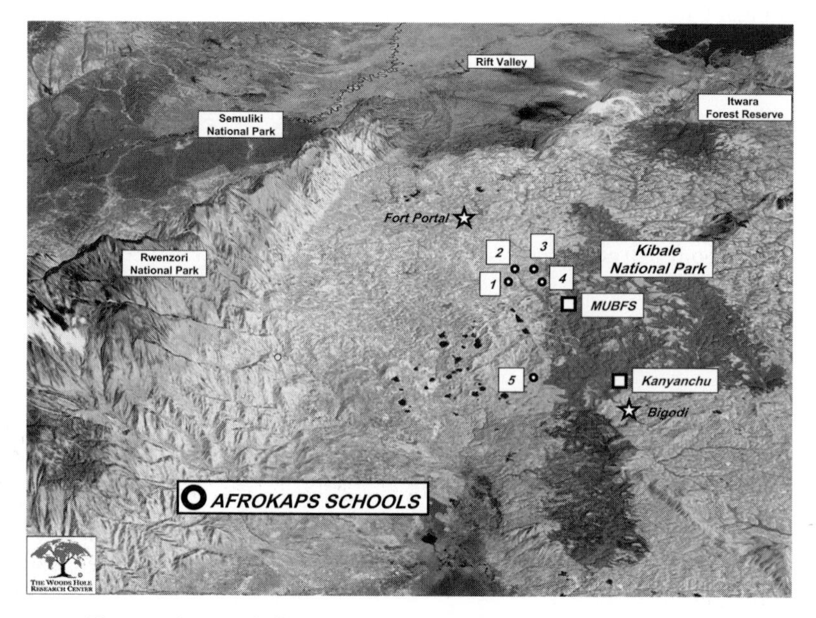

Fig. 10.2. A map of Kibale National Park showing the location of five schools supported by AFROKAPS and the Kasiisi Project.

universal primary education (UPE) and, for the first time, primary education was free. Overnight, the population at Kasiisi Primary School doubled from 429 to almost 900. The teachers, often unpaid for months, taught classes of up to 200 children in classrooms that were so unstable that they had to be evacuated during high winds. There was no place for teachers to meet and prepare lessons, no books, no paper, and no money for improvements. Drop-out rates were high. Participants in MUBFS, under the direction of John Kasenene, considered that poorly educated people were likely to turn to poaching to earn a living, and partly for this reason they decided to build a primary school at Kanyawara (school 4, Fig 10.2). Two classrooms and the local church formed the basis for the fledgling school. At the same time, a single visit to Kasiisi Primary School by Richard Wrangham, who had at that time been director of the Kibale Chimpanzee Project for 10 years, and his family alerted them to the dire conditions at that school. Partnering with John and Elizabeth Kasenene, a teacher at the school, the Wranghams started the Kasiisi Project, initially just to raise enough money to rebuild Kasiisi and to complete construction at Kanyawara. Since then, the project has expanded to include three other primary schools (Kiko, Kigarama, and Rweteera: schools 3, 2 and 5, Fig. 10.2) and, under the guidance of AFROKAPS since 1997, the Kasiisi Project has moved on from construction and now supports a variety of programs in the schools. The annual budget has grown from $1000 in 1997 to $70 000 in 2007, improving educational opportunities for 3500 children, up to 35% of whom are orphans.

Academic standards have risen substantially at Kasiisi. In 1997 no children at the school scored the top Grade 1 passes in their Primary Leaving Exams (PLE). By 2007, 25% of the graduating class had Grade 1 PLE passes. The project provides scholarships for bright, needy students, and in 2007 sponsored 52 graduates of AFROKAPS schools. In September 2007 Koojo Mathew will be the first graduate of the program to enter University.

The Kasiisi Project places a strong emphasis on educating girls. Women's education is well recognized as being critical for the future of developing countries. Educated women have smaller families, healthier children and contribute more to economic growth (King and Hill, 1993). The Kasiisi Project keeps girls in school by providing free sanitary towels, underwear, health education, and special latrines with incinerators and washing facilities. Preliminary data for 2005 and 2006 show that this intervention has reduced absenteeism in adolescent girls at Kanyawara School by 30%.

Other initiatives include conservation education, dormitories, latrines, libraries, staff rooms, clean water, workshops, and bursaries for teachers and nursery schools. Collaborating with Makerere University and the Presidential Initiative on Banana Industrial Development, we are launching a multi-year school lunch program at Kasiisi and Kanyawara schools in 2008, with the aim of improving attendance, nutritional status and academic success (Glewe *et al.*, 2001). Hoping to expand its impact, the project will help finance a Ugandan Ph.D. student whose study of the use of banana-based porridge to improve primary school nutrition is designed to be a model for school feeding interventions nationwide.

Books Open the World

This project was started by Gosia Arlet and Freerk Molleman in 2004. The project seeks to promote and encourage education and opportunity in rural communities for children and women through community libraries, literacy classes, vocational training, support groups, and by supporting all levels of education. Both directors have been involved in long-term research programs at MUBFS (on monkeys and butterflies, respectively, since 1999). Joel Hartter who works in the communities around KNP on landscapes around the park, joined the foundation in 2006. The money for the project is raised in the Netherlands and the USA to establish and support Community Libraries (now in five different villages), the tutor program in three primary schools, scholarships for postprimary education and literacy classes for adults (in five centers). The project also subsidizes the Kanyawara Nursery School and supports a local women's group with sewing machines, tailoring and English classes.

The Kasiisi Project is collaborating with this project by funding the purchase of secondary school textbooks for all the libraries.

The Kibale Student Support Program

The Kibale Student Support Program (KSSP) is a small-scale education project, which since 1988 has provided primary, secondary, and occasionally tertiary education for underprivileged Ugandan children, mainly orphans from poor families, living near the National Park. Children selected receive school fees, medical care, and life skill classes as well as uniforms, books, bicycles, sanitary supplies, etc. Since 1988, the project has supported a total of 51 children; 21 (9 boys and12 girls) in the 2007 school year. KSSP is directed by Marij Steenbeek, who, in cooperation with MUBFS, ran the Kibale Tree Planting Project, an environmental education project, from 1988 to 1993. Katooro Patrick and Dorothy Mbabazi, both former KSSP students, and teachers at Kasiisi and Kanyawara Primary

Schools supervise KSSP scholars. The project is supported chiefly by private donors in the Netherlands.

The Kasiisi Project and KSSP work together to support their secondary school students.

Kibale Community Fuel Wood Project

The Kibale Community Fuel Wood Project (KCFWP) was started in 2006 by Michael Stern and Rebecca Goldstone, who first came to KNP in 2000. Its goal is to provide an alternative source of wood to villagers who currently use Kibale National Park for fuel (Naughton-Treves, 1997), specifically by promoting fast-growing, indigenous leguminous trees. Wood and charcoal are the sole sources of energy for 98% of people surrounding KNP (MWLE, 2001; GoU, 1992) and even small-scale logging for fuel can severely damage the ecosystem (Bundestag, 1990). With a population growth of 2.6% (Spiegel, 2000) and the loss of forest fragments outside the park, which once were a source of fuel (personal communication, T. R. Gillespie, C. A. Chapman, and R. W. Wrangham) the forest is threatened.

In its pilot year with an investment of $39 203 the project created six demonstration areas, opened a natural history museum, hosted three workshops, presented 30 outdoor movie shows, and assisted in the building of over 100 fuel-efficient stoves (made for about one dollar each from local materials), engaging more than 11 000 villagers in all. Pre- and postsurveys will determine the effectiveness of these efforts and the next phase of the project.

This project is advising the Kasiisi Project on fuel wood cultivation and fuel-efficient stoves for their school-feeding program.

LESSONS FROM THE KNP EXPERIENCE

There are many factors that contribute to a successful project, which we are not going to discuss here: good accounting, supporting locally identified needs rather than imposing solutions from outside, contributions by the community, etc. What we want to describe here are the four things that we believe have been the cornerstones for the success of The Kasiisi Project. They are a consequence of the long-term research programs that the authors are connected to and we have seen that when they are in place the rest naturally follow. They are (a) good personal relationships founded on many years of familiarity, (b) frequent visits to cement these relationships, (c) a long-term commitment to the community, and (d) good data that show the results of interventions and inform plans for the future.

Personal relationships matter

Regardless of the kind of project, in our experience the best outcomes are reached when recipients and donors know each other well enough to understand and forgive some of the cultural pitfalls that are likely to cause problems. For example, we are aware that Ugandan men may find the assertiveness of western women disrespectful and in return western women, used to a more direct approach, are often impatient. The slower pace of life in Uganda can clash with the need of donors to balance their books annually, and erratic communication frustrates those who are used to rapid and prompt feedback. Coming from more affluent countries, expatriate donors often fail to appreciate the complexities of local political and financial pressures in developing countries. It is easy to underestimate the challenges faced by nationals trying to balance the expectations and needs of both donor and recipient cultures. All of these can hinder progress but our friendship and a common commitment to the communities we are trying to help mean that we can work amicably to solve the inevitable problems.

Frequent visits are vital

We cannot emphasize enough the importance of regular face-to-face contact. The foundations and charities that support us, responsible to their donors, and the US government, increasingly watchful of funds sent overseas, demand an honest assessment of the failures and successes of our project. On the other hand, recipients may fear that too much candor could lead to less generosity. We try to dispel the concern of all parties by frequent and regular visits by ER to Uganda to make sure that all parts of our project are focused on the same goal. Relaxed discussions and catching up on project news renew the ties that keep us on track. We find that restricting visits, despite their high cost, is a false economy. Fresh stories and up-to-date facts boost fundraising and donors, worried by stories of corruption, feel reassured when we can speak with the authority of personal experience.

AFROKAPS representatives visited the USA in 2007 and met our donor base. This helped the Ugandans understand some of the difficulties facing The Kasiisi Project as it tries to bridge the two cultures.

Long-term commitment pays off

For donors the problems are obvious, the giving relatively easy and providing assistance a pleasant option. For the Ugandans, however, there is grim need and no escape. They know that expatriates always leave in the end and, worried that they might lose money, they are likely to agree to projects even if they are doubtful of their chances of success. When it is clear that donors are committed for the long term, a more realistic assessment of needs and better allocation of funds follow.

With every passing year, the long-term health of The Kasiisi Project improves as we see the people who work for AFROKAPS become more invested in its success. The achievements of the project have become sources of personal pride. This has resulted in a subtle shift in attitudes at AFROKAPS. From being totally dependent on The Kasiisi Project for support they are now pro-active in soliciting funds from elsewhere. Since unexpected circumstances frequently intervene to end collaborations between nationals and expatriates, this kind of development is important as it fosters independence and helps prevent projects folding when the research program ends.

Good data are critical

Regular collection and analysis of relevant data have been important tools in ensuring the success of our project and one that comes naturally to scientists. We can assess where we are and plan for the future confident that we are basing our decisions on facts. Accurate records of staff and student attendance, at AFROKAPS supported schools showed us that high staff absenteeism is a major contributor to poor academic performance as measured by Primary Leaving Exam results (Fig. 10.3). Other possible factors, including amount of money invested, percentage of children in single parent families, and differences in staff to student ratios, did not correlate with academic performance. This information will allow us to design interventions most likely to have the impacts we are aiming for.

SUMMARY

Pure research, even though it does not appear to directly impact people's lives, can have a profound effect on the socioeconomic status of communities around established research sites, while at the same time helping conserve vital ecosystems. An example is the Kibale Chimpanzee Project, which has been doing research on chimpanzee behavior in KNP for the past 20 years. This long-term research program led to the development of ecotourism at Kanyanchu with all its spin-offs and benefits for the Bigodi community, the support of local schools through The Kasiisi Project

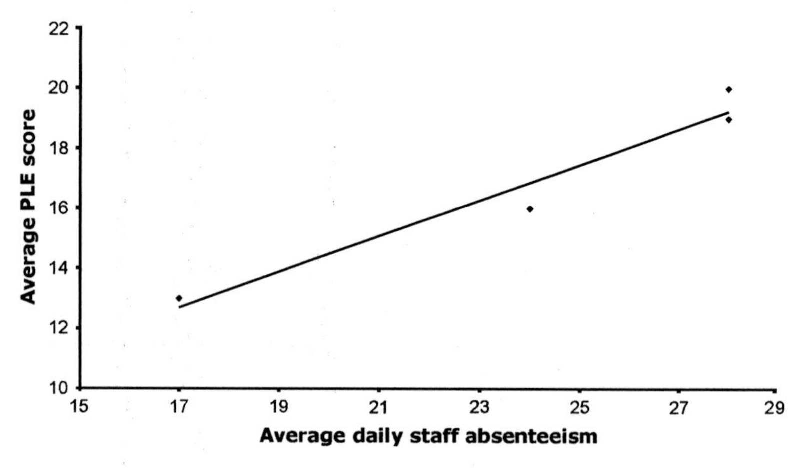

Fig. 10.3. Graph showing the correlation between Primary Leaving Exam (PLE) results and staff absenteeism in three primary schools in Kibale National Park. PLE results improve (scores are lower) as staff absenteeism declines.

and the establishment of the Kibale Snare Removal Project (Wrangham, Chapter 1).

Our experience in Uganda is that the expatriate research community represents an enthusiastic source of support that has led to helpful projects with apparent long-term benefits for conservation. Small projects, with low overheads, can give very good value for money, and the familiarity of the project directors with the sponsored communities gives confidence to donors who like the personal connection. Successful projects have a better chance of engendering positive attitudes to conservation.

But investors in projects have sometimes been naïve about differences in culture that have led to project failure. Such difficulties have been overcome by the development of alliances based on a clear-sighted understanding of, sometimes competing, personal and political pressures on both donors and recipients. This takes long-term investment, personal friendships and experience-based awareness of the differences between expatriates and nationals in goals and management styles. Long-term research programs play a critical role in the success of such ventures, providing the environment that both nurtures the birth of community projects and ensures that they continue to benefit the local people.

It is hard to measure whether the kind of community projects that have arisen as a consequence of long-term research programs in KNP have really made the local people want to protect the forest and its resources.

Struhsaker and colleagues (Struhsaker *et al.*, 2005) were unable to find a positive correlation between community investment and conservation. However, they did find that a positive attitude of communities towards Protected Areas was the strongest correlate of Protected Area success. Goldman *et al.* (Chapter 12) have data that show a positive attitude to Kibale National Park in the people they questioned. This attitude to the Park is repeated in an informal questionnaire completed by pupils of Kasiisi and Kanyawara schools in their villages, where 80% of those who approved of the National Park also said that they thought that the Kasiisi Project was good for the community. It is tempting to connect the dots.

ACKNOWLEDGMENTS

We would like to acknowledge the following for their assistance in writing this chapter and their support of our project: The Uganda Wildlife Authority, the Chief Park Warden KOA, MUBFS, The Kabarole District Education Office, Gosia Arlet, Rebecca Goldstone, Beatrice Kabahuma, Joshua Kagaba, Sonya Kahlenberg, Isaac Kaihura, Elizabeth Kasenene, Innocent Kato, Wayne Lobb, Moses Mapesa, Freerk Molleman, Arthur Mugisha, Martin Muller, Florence Muranga, Lisa Naughton, Brian O'Connor, Emily Otali, Marij Steenbeck, Michael Stern, Barbara Stevens, Charles Tumwesigye, Richard Wrangham and the parents, teachers and pupils of Kasiisi, Kanyawara, Kigerama, Kiiko and Rweteera primary schools. We also thank the generous donors to all the community projects around Kibale National Park, especially the congregation of First Parish Church in Weston and the Weston Public Schools for their loyal support of the Kasiisi Project. Agencies, individuals, and foundations underwriting the projects described in this chapter can be accessed on their websites.

REFERENCES

Baranga, D. (1983). Phenological observation of two food tree species of colobus monkeys. *African Journal of Ecology*, **24**, 2089–2214.

Basuta, I. G. (1987). The ecology and conservation status of the Chimpanzee (*Pan troglodytes* Blumenbach) in Kibale Forest, Uganda. Ph.D. thesis, Makerere University, Kampala, Uganda.

Buhapa, E. (1994). Elephant ecology in Kibale National Park and its peripherals. M.Sc. thesis Makerere University, Kampala.

Bundestag (ed.) (1990). *Protecting the Tropical Forests: A High Priority International Task.* Bonn: Bonner Universitats-Buchdruckerei.

Butynski, T. M. (1982). Harem – male replacement and infanticide in the blue monkey (*Cercopithecus mitis stuhlmanii*) in the Kibale Forest, Uganda. *American Journal of Primatology*, **3**, 1–22.

Chapman, C. A. and Chapman, L. J. (1996). Exotic tree plantations and the rehabilitation of natural forests in Kibale Forest National Park. Uganda. *Biological Conservation*, **76**, 253–257.

Glewe, P., Jacoby, H., and King, E. (2001). Early childhood nutrition and academic achievement: a longitudinal analysis. FCND Discussion Paper, 68.

Government of Uganda (GoU) (1992). *1991 National Housing and Rural Settlements Census*. Kampala, Uganda.

Kakudidi, E., Bukenya-Ziraba, R., and Kasenene, J. M. (2000). The medicinal plants in and around Kibale National Park. *Uganda Lidia*, **5**, 109–124.

Kapiriri, M. N. (1980). Non-timber forest products of Kibale Forest: their contribution to the local subsistence. M.Sc. thesis, Makerere University.

Kasenene, J. M. (1984). The influence of selective logging on rodent populations and the regeneration of selected tree species in the Kibale Forest, Uganda. *Tropical Ecology*, **25**, 179–195.

Kasenene, J. M. (1998). Forest association and phenology of wild coffee in Kibale National Park. *African Journal of Ecology*, **36**, 241–250.

King, E. M. and Hill, M. A. (1993). *Women's Education in Developing Countries; Barriers, Benefits and Policies*. World Bank Publications.

MWLE; Ministry for Water, Lands and Environment (2001). The Uganda Forestry Policy. Kampala, 29.

Naughton-Treves, L. (1997) Farming the forest edge: vulnerable places and people around Kibale National Park, Uganda. *Geographical Review*, **87**, 27–46.

Naughton-Treves, L. (1998). Predicting patterns of crop damage by wildlife around Kibale National Park, Uganda. *Conservation Biology*, **1**, 156–168.

Oates, J. F. (1974). The ecology and behavior of the black and white colobus monkey (*Colobus guereza* Rueppel) in East Africa. Ph.D. thesis, University of London, London, UK.

Rudran, R. (1978). Socioecology of blue monkeys (*Cercopithecus mitis, stuhlmanni*) in the Kibale Forest, Uganda. *Smithsonian Contribution to Zoology*, 249.

Spiegel Almanach Weltjahrbuch (2000). Spiegel Buchverlag, Hamburg, WEC World Energy Council 2001. In *Survey of Energy Sources*, ed. J. Trinnaman and A. Clarke, pp. 638.

Struhsaker, T. T. (1975). *The Red Colobus Monkey*. Chicago: University of Chicago Press.

Struhsaker, T. T., Struhsaker, P. J., and Siex, K. S. (2005). Conserving Africa's rain forests: problems in protected areas and possible solutions. *Biological Conservation*, **123**, 45–54.

Van Orsdol, K. (1986). Agricultural encroachment in Uganda's Kibale Forest. *Oryx*, **22**, 115–118.

ARTHUR MUGISHA

11

Potential interactions of research with the development and management of ecotourism

INTRODUCTION

Wildlife-related research is the foundation of wildlife management and especially of tourism development. Research provides vital information that generates, in potential tourists, an interest in and a desire to see wildlife, and adds value to the tourism products offered by wildlife managers and tour guides.

Ecotourism has come to be an illusive term. It is often so loosely applied that, at times, it can mean everything or nothing about the tourist experience on offer. In this chapter ecotourism is defined as that kind of tourism which is nature-based but which is also designed to impart understanding and learning about the culture and history of a given environment. Local communities benefit while at the same time care is taken not to alter the integrity of the ecosystems.

Based on this definition of ecotourism, I argue that to understand and learn more about a given ecosystem, long-term research is essential. This is particularly crucial for the understanding of those natural processes that take a long time to manifest themselves. I also point out that particularly valuable tourist sites, such as those that allow visitors to see primates, have followed in the wake of long-term research programs.

It is commonly believed among local people in Uganda that big mammals such as chimpanzees (*Pan troglodytes*) and gorillas (*Gorilla gorilla*) are dangerous and therefore should be killed. However, due to the information about these animals resulting from established research projects and the educational outreach that accompanies them, attitudes based on such false beliefs have changed. Local communities now welcome the habituation programs that have become the key to primate ecotourism. The success of these sites and the conservation of their ecosystems is dependent on the information that comes from continuing research. The challenges facing wildlife managers, for example, disease transmission between primates and people (Goldberg *et al.*, Chapter 8), and the possibility that the behaviors of the primates may be affected by exposure to tourists negatively impacting their health and population status, need long-term data to resolve them. However, the financial and economic gains from the ecotourism industry are so tempting that such challenges are often considered trivial and ignored. We need accurate long-term researched information to assist policy makers and to ensure that we adhere to the conservation principles of a well-managed ecotourism industry. Also, we need long-term research to understand both what attracts people to ecotourism, and the changing trends in what is popular so that we can plan for, and provide, a satisfying experience for our visitors.

Furthermore, there are other challenges that affect the wider ecosystem such as global warming. We need data to predict relationships between human activities and the welfare of the natural resource base. This calls for more investment in long-term research and the promotion of policy formulation and program implementation that is based on accurate researched data.

RESEARCH IN KIBALE

The present-day Kibale National Park was gazetted in 1993. Before it was declared a National Park, it was divided into two parts which were managed separately. The Kibale Forest Reserve, declared in 1948, and the Kibale Forest Game Corridor, declared in 1964, were managed by two different Government institutions, the Forest and the Game Departments, respectively. The management objectives of these two departments were different. The Game Department managed the Game Corridor to control the seasonal movement of animals, especially elephants, between the Kibale Central Forest Reserve to the north and the Queen Elizabeth National Park to the south. The management objective of the Forest Reserve was to produce timber on a commercial basis. There was little or no interest in biodiversity conservation by either organization.

In 1970, the Kibale Forest Project was founded. Its initial objective was to study the behavior and ecology of the endangered Uganda red colobus (*Procolobus rufomitratus*) (Struhsaker, 1997). This study formed the basis of nearly 40 years of continuous research into the forest's animals and plants. By describing the biodiversity of Kibale forest and by identifying the need to conserve it, this research ultimately led to a movement to upgrade the conservation status of the forest from a Reserve to a National Park (Struhsaker, Chapter 4).

The research that started as the Kibale Forest Project provided not only the scientific information that highlighted the need for biodiversity conservation, but also challenged the management objective of the Forest Department, which was initially commercial timber production. However, even before Kibale was officially declared a National Park in 1993, the Forest Department had already gone some way towards changing the emphasis of its management practices from timber production to biodiversity conservation and the development of tourist facilities. In 1992, a year before the reserve was declared a National Park, a tourist center was started at Kanyanchu. This shift in the management emphasis was due largely to the establishment of the Makerere University Biological Field Station (MUBFS). Begun in 1987, MUBFS had by this time gained popularity amongst donor and international communities as a credible research station. The short-term biodiversity surveys undertaken by Frontiers, a United Kingdom-based non-governmental organization (NGO) and the Forest Department supported by the European Union, added to the understanding of the rich diversity of Kibale Forest. Despite this progress, the relationship between the research station and forest management and between forest management and researchers often ran into difficulties. The problems that existed then, between research and management in Kibale National Park, symbolize the sometimes-conflicting priorities and values of the different constituencies involved in the management of natural resources. For example, when the Reserve was declared a National Park, the relationship between management (represented by the National Parks) and research (represented by MUBFS) was never properly defined, with the result that the role of research in the newly established Kanyanchu tourist center was never a priority. Increasing revenues from Kanyanchu also highlighted differences in priorities for the two groups. Whereas the Park management wanted to maximize tourist revenues, the researchers were more interested in the long-term impacts of the tourism industry on the forest ecosystem. However, it is only when Protected Area authorities begin to understand and appreciate the role of research in the management of National Parks and Reserves that we are able to take advantage of the benefits of long-term data and use them to guard our natural resources. The experience in Kibale National Park has been that researchers have continued to be interested in the tourism development taking place in Kanyanchu and have actively lobbied the management to take account of their research findings. This greatly influenced the direction in which ecotourism at Kanyanchu developed.

THE ROLE OF RESEARCH IN DEVELOPING THE ECOTOURISM INDUSTRY IN KIBALE NATIONAL PARK

Before I go further in discussing the interactions between research and ecotourism development, let me take a step back to define what I mean by ecotourism in more detail. The term ecotourism is used increasingly commonly in the tourism fraternity. The origins of this term can be traced to the negative impacts of the fast growth of tourism on key sites that were also biodiversity hotspots and home to indigenous people. For instance, inadequate disposal of untreated sewage has damaged critical habitats important for biodiversity conservation. Greater numbers of tourists require more energy and water, which can strain local resources and damage the environment. Furthermore, critics of tourism are concerned about the development of additional attractions, e.g., golf courses, which lead to soil and water pollution and degradation of biological resources. The social and cultural aspects of a site can be equally impacted negatively by tourism, especially where economies are developed through the promotion of foreign cultures at the expense of local indigenous ones. Moreover, critics further claim that tourism aggravates social imbalances through the promotion of such activities as the sex trade and child prostitution and labor. They further argue that the local population gains little from tourism and foreign earnings, if any, are used to meet increased demands for imported products in a further bid to attract more tourists (Hausler in Mugisha and Ajarova, 2006).

It was out of such concerns that conservationists coined the term ecotourism. The central idea was to take into account the impact that tourism can have on the environment as well as on the lives of local people and its potential to destroy the very thing that attracted the tourists in the first place. However, I think this idea has been somehow "hijacked" by those that put financial interests at the center of their programs. The term is currently being used as marketing jargon and may convey little about the true nature of the product being offered. In Mugisha and Ajarova (2006), ecotourism is defined as a "type of tourism that aims at promoting purposeful travel to natural areas to understand the culture and history of the environment, taking care not to alter the integrity of the ecosystem, while producing economic opportunities that make conservation of natural resources beneficial to local people." It is this type of tourism that I refer to when discussing the importance, in this chapter, of long-term research for ecotourism.

The key attractions that promoted Kibale National Park as a tourist destination were the number and density of primate species, especially chimpanzees, in the forest. In the early 1990s the Kibale Chimpanzee Project (KCP), begun when Kibale was a Forest Reserve, was operating under the auspices of MUBFS in Kanyawara. It was committed to a long-term research program to study the behavioral ecology of a community of chimpanzees whose home range included the forest around the field station. The interest that was exhibited in this habituated research group of chimpanzees by non-researchers and the subsequent pressures on the Kanyawara community underscored the need to identify and habituate a separate chimpanzee community for tourists. This meshed with the idea that an important way to contribute to the conservation of Kibale was to develop a chimpanzee ecotourism project. Richard Wrangham, director of KCP, wrote a proposal for an ecotourism project and the Wildlife Conservation Society commissioned a feasibility report. Access to a second group of chimpanzees also provided an opportunity for interesting comparative studies between the two communities (R. Wrangham, personal communication). Habituation of a community of chimpanzees, specifically for tourism, was begun at Kanyanchu in the forest south of the field station.

After the need to establish a tourist center was identified and the center developed, research continued to provide much needed data on the impacts of this industry on the ecosystem and helped define the product that would be offered. For example, research carried out by Obua and Harding (1997), showed that a substantial amount of damage was occurring at campsites in Kanyanchu, despite the low number of tourists visiting the park (5000 in 1996). This was small compared to tourist records from other National Parks in Uganda at that time, so such research was important for the developing site, to ensure a product that would not negatively impact the environment. This is an example of how research can interact with management to shape the way tourism in Protected Areas is developed.

ECOTOURISM IN KIBALE NATIONAL PARK

The Kanyanchu tourist center is located in the southeastern part of Kibale National Park, a few meters off the Fort-Portal to Bigodi road (Fig. 11.1). As noted above, it was designed to provide an environmentally friendly tourism product that would allow visitors to see chimpanzees. It

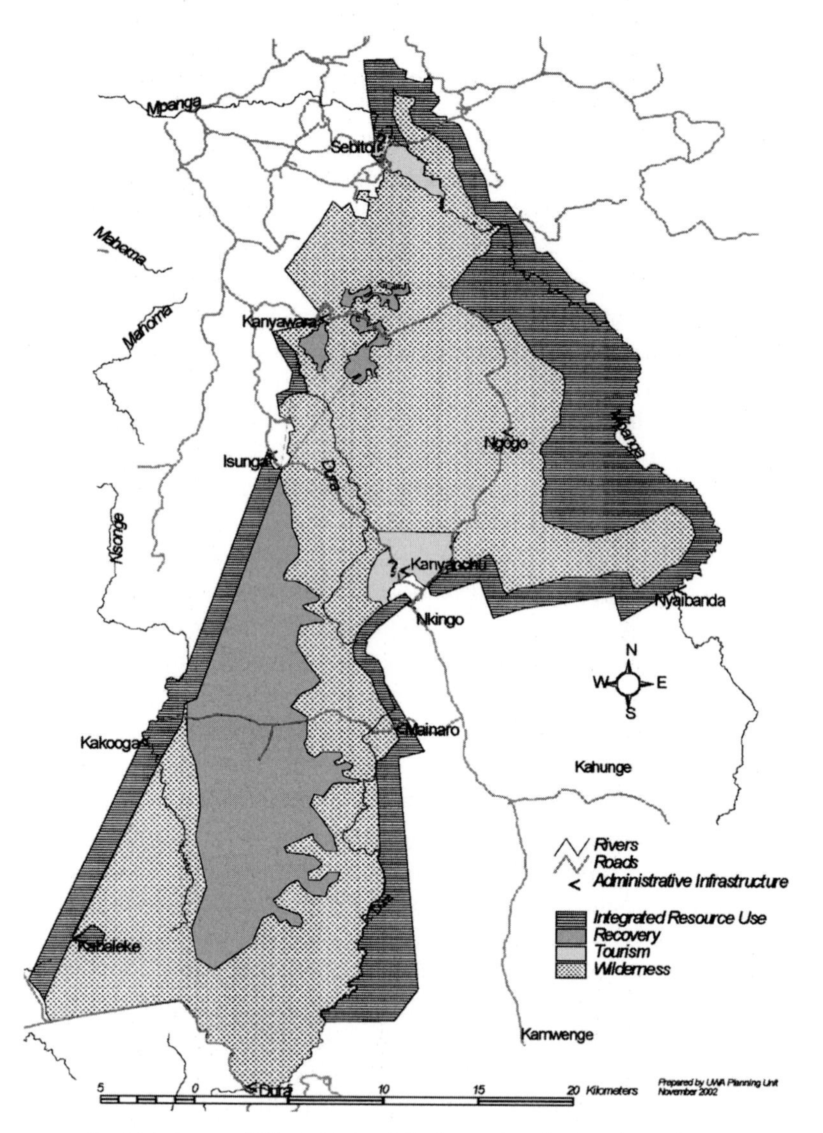

Fig. 11.1. A map of Kibale National Park showing the locations of Kanyawara (Makerere University Biological Field Station) and Kanyanchu (Ecotourism Project).

was hoped that, if a tourist site were to be developed, awareness of the importance of conserving Kibale National Park would increase at local, national, and international levels. In addition, ecotourism had the potential to become one way to replace incomes from traditional timber extraction, lost when the Reserve became a National Park. It might also go some way towards compensating local communities for the loss of access to forest resources as well as improving relations between them and the National Park.

Kibale Forest Reserve was previously recognized for harvesting of timber, and the history of the logging industry in Kibale is well documented (Struhsaker, 1997). Although logging was a big industry, it did not deliberately target local communities as beneficiaries. In fact, the major way in which the communities benefited, if at all, was through access to local employment opportunities in the softwood forest plantations that were started to boost timber production. Another benefit to the communities during this time that must be mentioned, was unrestricted access to the natural resources within the reserve. Law enforcement was not strong enough to prevent such activities, and communities neighboring the forest were able to harvest forest products ranging from building poles to bushmeat by illegal hunting (Struhsaker, 1997).

With the upgrading of the conservation status from a Forest Reserve to a National Park, there were serious concerns from neighboring communities that people's livelihoods would be negatively impacted. These negative attitudes towards the Park increased with allegations that the Government had imported crop-raiding and man-eating wildlife into the forest (Mugisha, 2002). With support from conservation organizations, such as the International Union for the Conservation of Nature (IUCN), interventions were devised to improve community attitudes by integrating conservation and community development needs. Although there were significant achievements from these efforts, people still harbored resentful feelings towards the Park. This was partly due to the fact that the effects of occasional crop raiding were felt at a household or individual level, whereas the conservation and development project benefits were realized at a community level.

Ecotourism began in Kibale National Park in 1992, after the Kibale Forest Reserve was converted into a Forest Park and declared a multiple use Protected Area. The first tourists were received in January 1992 and used trails in the Kanyanchu grasslands designed by the British-based NGO, Frontiers (Obua and Harding, 1997). The first amenities developed were the camping site and information center along with basic staff accommodation. More than 80% of the staff were resident in Bigodi, the nearby village. Guides and support staff cycled or walked every morning from their homes to Kanyanchu for work. Being able to see the wildlife in the forest was the main draw for tourists visiting Kanyachu, with chimpanzee tracking being the major attraction. In order to provide a good experience for the tourists, well-trained guides were needed. Initially, this training was done by KCP, which was able to supply relevant information and advice based on its own long-term research findings. Frontiers continued with this training in collaboration with KCP. KCP also hosted seminars for trainee guides and sent its own field assistants to Kanyanchu to help with training in the forest. It is important to note that, initially, it was these research assistants using experience with KCP who pioneered chimpanzee tracking at Kanyanchu. Their tracking skills and knowledge of chimpanzee behavior was transferred to tour guides who then trained others.

With the commencement of the ecotourism industry in Kanyanchu, the Park's management gained another important "tool" for involving the neighboring communities in managing the protected area and winning them over to the benefits of conservation. The neighboring community of Bigodi took a lead in getting fully engaged in the management of the tourism industry. Through the training they received from Frontiers, access to research findings and help with capacity building from the Kibale Forest Project, the Bigodi residents began to involve themselves in the tourism business. With more support from the Peace Corp Volunteer Program, starting 2 years after the commencement of the ecotourism services at Kanyanchu, tourist services began to develop spontaneously in Bigodi. First were eating places targeting day visitors to Kanyanchu. This was followed later by accommodation that catered mainly for opportunistic visitors who visited Kanyanchu on their own, without pre-arrangements. Tour operators used camping sites in Kanyanchu or outside the Park. Bigodi women organized themselves into a cooperative group, established a kiosk in Kanyanchu and began selling their handcrafts and other products. Increasing numbers of tourists began arriving in Kanyanchu and, with the Park's restrictions on both the numbers of people permitted to track chimpanzees and the duration of these treks, demand soon outstripped supply. Recognizing the need for more tourist products, the Bigodi community once again responded. This time, the community took a big leap forward and developed the Magombe Swamp as an alternative tourist destination. This swamp, near Bigodi, lies outside the park and offers bird-watching walks and the opportunity to see primates other than chimpanzees. This initiative was possible (a) because of the involvement of trained guides who had learnt their trade in

Kanyanchu and (b) because of the ability of the community to organize it. They formed the Kibale Association for Rural and Environment Development (KAFRED). These developments contributed significantly to the successful management of the natural resources of Kibale National Park at a time when there were fears that the approach of Uganda National Parks (later the Uganda Wildlife Authority) to managing the forests would clash with the interests of the local people.

These developments helped the National Park authorities justify the existence of Kibale National Park to the community, and greatly assisted in developing a positive attitude to the park. For example, Lepp (2007) reported that residents generally react favorably to tourism in Bigodi. This results from the belief that tourism creates community development, improves agricultural markets, generates income, and brings random good fortune through exposure to the international community. The success of ecotourism at Kanyanchu and its multiple beneficial impacts on the local community can be attributed to management practices that grew out of the foundation of long-term research. The gazetting of Kibale as a National Park restricted the access of local people to natural resources in the forest but ecotourism provided alternative income and assisted in empowering the surrounding communities and genuinely involving them in park management.

THE ECOTOURISM PRODUCT IN KIBALE NATIONAL PARK

Chimpanzee sightings

To ensure that visitors to Kibale leave satisfied by their experiences in the Park, the success rate of encountering the habituated groups of chimpanzees on every visit needed to be increased. It was therefore decided that habituation of the tourist community at Kanyanchu had to be improved and a number of staff were allocated solely to this task. With increasing tourist arrivals signifying greater interest in the ecotourism industry in Kibale, the Park's management decided to develop a general management plan. This plan was designed to guide and direct tourism development in Kibale National Park. Under the tourism plan, an "all-day chimpanzee follow" option was offered. This was designed to help tourists understand how the chimpanzee families spent their days in the forest, from the time they "got up" from their nests in the morning to the time they "nested" again in the evening. Tourist numbers have been on the increase since then, from about 1300 arrivals per annum in 1992 to more than 6000 tourists in 2006 (Table 11.1).

Table 11.1. *Number of tourists visting Kibale Forest 1992–2006*

Year	1992	1993	1994	1995	1996	1997	1998	1999	2000	2001	2002	2003	2004	2005	2006
Number of visitors	1297	3026	3167	3304	4017	2449	2003	955	1149	1846	4899	5998	5463	6490	6960

Source: UWA Records.

In addition to the chimpanzee tracking and to ensure a diversified product, which would reduce pressure on chimpanzees, other options for tourists have been developed over time. These include nature trails, bird watching, red colobus monkey trekking, and night-time long distance nature walks. The development of these alternatives is based on the need to reduce the negative impacts that providing only one option can have and also to spread possible benefits to as many of the communities neighboring the Park as possible. This planning is informed by long-term research data and will ensure that the ecotourism product is kept competitive.

Accommodation facilities in Kibale National Park

Other services that define the ecotourism product include accommodation facilities. At Kanyanchu, the first accommodation was tents inherited from Frontiers that provided basic shelter. Management then started marketing their product to tour operators, who provided professionally organized tented safaris for their clients, in designated camping sites. Later, management built small bandas equipped with locally designed warm water showers and basic bedding. As national security continues to improve in Uganda and the tourism industry to grow, a number of other facilities developed by the private sector are springing up outside the Park. These will further enable the neighboring communities to benefit from the ecotourism, increasing support for Kibale National Park and the management plans that conserve its biodiversity.

CHALLENGES FACING ECOTOURISM IN KIBALE NATIONAL PARK

Relations between long-term researchers and natural resources managers

As discussed above, for ecotourism to meet the principles and expectations of our definition of ecotourism, there is a need for continued research and monitoring with the research data being fed into the management decision-making system. However, the results of long-term research, by definition, are not immediately available and often these results are confounded by other factors so it can be difficult to clearly relate cause and effect. Managers, on the other hand, are frequently looking for direct and readily available solutions to management problems. This can cause difficult relationships between long-term research programs and natural resources managers. Managers end up accusing researchers of wasting time on research that is too academic and the researchers accuse the managers of not respecting their research efforts. Such challenges were witnessed in the research history of Kibale National Park. However, with continued interactions, both informal and formal, there has been a reduction in these problems with time.

Threat of disease transmission between humans and primates

One of the emerging concerns that will have a big impact on the ecotourism industry, especially where primates are concerned, is the risk of disease transmission between human beings and primates (Goldberg *et al.*, Chapter 8). In the Democratic Republic of the Congo (DRC) and Gabon for example, it is recorded that more than 5000 western gorillas (*Gorilla gorilla gorilla*) and chimpanzees were decimated by Ebola virus over the past decade (Bermejo *et al.*, 2006). The reservoir for the Ebola virus was probably fruit bats but there may have been transmission from apes to humans (Leroy *et al.*, 2005).

Other viral diseases such as bird flu are becoming a real menace in Africa and even beyond. At the time when the chimpanzee habituation process began at Kanyanchu, it was assumed that the rules and regulations governing the visitations to these endangered species would protect them from disease threats. Regulations stipulated the minimum distance between the visitors and the primates, the health status of the visitors, the generation and disposal of waste, the number of visitors per group, and the length of visits, among other things. However, over time and as the primate trekking has become more popular and an increasingly important source of revenue for the Protected Areas, it has become clear that the enforcement of these regulations leaves a lot to be desired. This makes the primates vulnerable. Sandbrook and Semple (2006), in their research in Bwindi Impenetrable National Park, found that the rules are not being adhered to. The mean distance observed between gorillas and humans was recorded to be 2.76 meters as opposed to the regulated 7.0 meters and this contact was observed lasting for more than 1 hour. Managers and policy makers often fail to appreciate the need to observe such regulations and, instead, consider maximizing the income generated from ecotourism programs as their primary goal.

On the need to limit the number of visitors visiting a group, they often argue that primates do not know how to count, and use such arguments to increase the number in a visiting group. On the ground, the tour guides often fail to appreciate the need to keep to the regulated minimum distance and length of stay and are frequently ready to compromise on these regulations in anticipation of a better tip. The irony is that such complacency, about what are perceived to be trivial issues, can lead to the destruction of the tourism and its revenue. For example, the gorilla census that was carried out in Bwindi National Park in Uganda in 2002 showed that the gorilla population distribution in the Park was affected by human disturbance. Although there was a noted increase in the overall population, the ratio of young to adults remained the same (McNeilage and Robbins, 2006), indicating a population structure that is unstable in the long run. A desired situation would be where the ratio of the young adults to adults is higher than the ratio of adults to young, and is increasing. These challenges need to be addressed by using well-researched and accurate information and by continuous engagement between researchers and natural resources managers. Also, in light of the fact that the enforcement of formulated regulations is difficult, research needs to take a lead in recommending other options such as the need for disinfected face-masks to address threats of infection.

Climatic change

Another challenge that needs to be addressed through long-term research is the ongoing concern about climate change and global warming. It is anticipated that the health of the ecosystem that forms part of the primates' home in Uganda will be impacted by climate change and global warming. It is not yet clear what the implications of such changes will be on the behavior of the primates. Long-term research will be instrumental in providing good data which will increase the accuracy of any predictions of the impacts of this phenomenon on the health of the ecosystem as well as on the behavior of the primates.

SUMMARY

Research and management of natural resources, especially in establishing ecotourism programs, should be twin programs to ensure successful intervention. However, due to the need for short-term, urgent solutions, managers at times become impatient and lose interest in research findings. It is important that researchers take deliberate efforts to continue engaging managers in their research activities. In this way, the chances of management taking future research into account will be optimized.

The ecotourism program in Kanyanchu was a direct result of long-term research by the Kibale Chimpanzee Project. In fact, the upgrading of the conservation status of the Kibale Forest Reserve to that of a National Park was itself a direct result of the long-term research of the Kibale Forest Project that was started in the 1970s. This research has continued to inform and shape the development of ecotourism and its benefits have spread to the neighboring communities and the private sector. However, there is a need for increased and continuous research to address the emerging challenges that are confronting the ecotourism programs in Kibale National Park. As McNeilage and Robbins *et al.* (2006) put it, "scientific research . . . is a global endeavor which increases our understanding of the world around us and how it functions." This understanding is critical for conservation.

REFERENCES

Bermejo, M., Rodríguez-Teijeiro, J. D., Illera, G., Barroso, A., Vilà, C., and Walsh, P. D. (2006). Ebola outbreak killed 5000 gorillas. *Science*, **314**, 1564.

Lepp, A. (2007). Residents' attitudes towards tourism in Bigodi village, Uganda. *Tourism Management*, **28**, 876–885.

Leroy, E. M., Kumulungui, B., Poumut, X., Rouquet, P., Hassanin, A., and Yaba, P. (2005). Fruit bats as reservoirs of Ebola virus. *Nature*, **438**, 575–576.

McNeilage, A. and Robbins, M. M. (2006). Primatology comes to Africa. *African Journal of Ecology*, **45**, 1–3.

McNeilage, A., Robbins, M. M., Gray, M. *et al.* (2006). Census of the mountain gorilla *Gorilla beringei beringei* population in Bwindi Impenetrable National Park, Uganda. *Oryx*, **40**, 419–427.

Mugisha, R. A. (2002). Evaluation of community based conservation approaches. Management of protected areas in Uganda. Dissertation, University of Florida.

Mugisha, A. and Ajarova, L. (2006). Ecotourism: benefits and challenges – Uganda's experience. In *Gaining Ground: In Pursuit of Ecological Sustainability*, ed. D. Lavigne. International Fund for International Welfare and the University of Limerick, pp. 153–159.

Obua, J. and Harding, D. M. (1997). Environmental impact of eco-tourism in Kibale National Park, Uganda. *Journal of Sustainable Tourism*, **5**, 213–223.

Sandbrook, C. and Semple, S. (2006). The rules and the reality of mountain gorilla – *Gorilla beringei beringei* tracking: how close do tourists get? *Oryx*, **40**, 428–433.

Struhsaker, T. T. (1997). *Ecology of an African Rainforest. Logging in Kibale and the Conflict between Conservation and Exploitation*. Gainsville: University Press of Florida.

ABE GOLDMAN, JOEL HARTTER, JANE SOUTHWORTH,
AND MICHAEL BINFORD

12

The human landscape around the Island Park: impacts and responses to Kibale National Park

Most social scientists in recent years have written about parks very differently from biologists and other promoters of Protected Area conservation. Especially when dealing with Africa and other developing regions, social scientists have generally portrayed parks as areas of restriction and exclusion imposed on a disempowered poor rural population by the combined forces of national governments and international conservation movements (Brockington, 2002; Gibson, 1999; Guyer and Richards, 1996; Neumann, 1998, 2001). On the whole, local people are seen to have little say and to derive little or no benefit from parks that neighbor them and that in many cases occupy land they previously controlled. Indeed, they often are not even permitted within the boundaries of the parks, and activities they might have performed routinely in the past have, in the context of the park, become illegal and subject to severe penalties. People living near parks also face the risks of wild fauna, which in many cases threaten their lives, livelihoods, and property – risks that few people in rich countries would tolerate for long.

The main counter examples to such negative depictions of "fortress conservation" have been the portrayals of community conservation programs, in which local communities are given a direct stake in the preservation of habitats and animal populations near them. In most of these cases, local people are involved in management decisions and have control over a substantial portion of the revenue collected by the protected area. Most examples of relatively successful community-centered park conservation have been in regions of southern Africa, and their main financial basis has been the sale of hunting licenses (Child, 2004; Fabricius and Koch, 2004; Hulme and Murphree, 2001). There have been few similarly successful examples in eastern Africa, or other regions of sub-Saharan Africa. There is a range of possible explanations for the geographic concentration of successful community park conservation which will be touched on (though not dealt with exhaustively) below.

PROJECT DESCRIPTION

This chapter discusses some of the findings of an interdisciplinary study that examines how parks in eastern Africa have affected land use, livelihoods, and biodiversity in the landscapes surrounding them. The project addresses two basic research questions: (1) How does the presence of a park affect agricultural land use and other livelihood strategies by the people living near the park? (2) How does the character of agriculture around the park affect biodiversity outside the park? These questions are investigated through analysis of land cover change over several decades based on satellite imagery; social science research involving surveys and interviews with people living around the park as well as park managers and policy makers; and biological sampling of key indicator species outside the park. Among other factors, we have been investigating the degree to which distance from the park boundaries affects people's livelihood and risk behavior as well as biodiversity conditions as a way to assess the impacts of the park. In addition to Kibale National Park (KNP), the first phase of the study has included Tarangire National Park in northern Tanzania. This chapter deals only with some of the social science research around KNP.

One of the main features of this project has been our use of random spatial sampling for interdisciplinary research. Given the population and intensive land use conditions around KNP, we have focused on an area within 5 kilometers of the park boundary to study the park's impacts on the surrounding landscape. (In a less densely populated region, a larger buffer zone would generally be needed.) We have identified two research areas on the east and west sides of the park that differ to some extent in land use, ethnicity, infrastructure, and altitude (see Fig. 12.1). We selected a set of random geographic coordinates within these areas, and those points became the centers of 9-hectare areas (circles with radius of about 170 meters) that we term "superpixels." These have been the basis for most of our social science and biological sampling and data collection and some of our land cover analysis. We ultimately used 95 superpixels, 60 on the west and 35 on the east side of the park (the proportions result

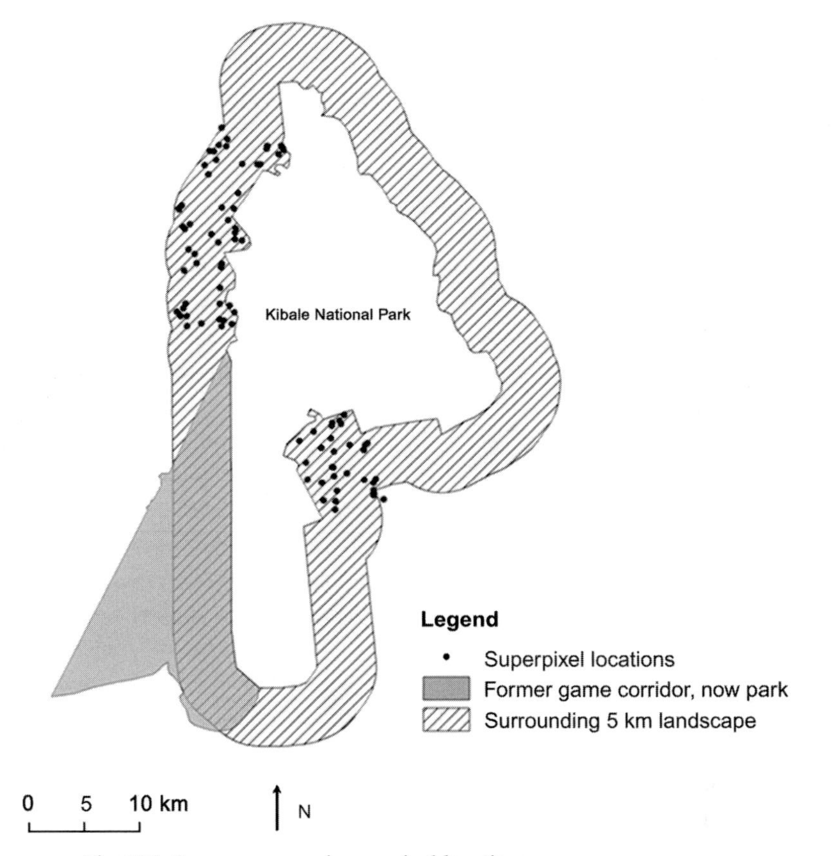

Fig. 12.1. Survey areas and superpixel locations.

from random selection and the respective sizes of the two areas). In our social science research, we have interviewed one or more people who own or use land in each superpixel. Individual superpixels can include anywhere from zero or one to several dozen land holders. (Some superpixels are entirely within tea estates or lakes and thus have no individual land owner[1]. At the other extreme, one of the superpixels fell in the center of a village.)

This sampling scheme gives us a representative sample of land use conditions in the survey regions, and it differs from most purely social scientific samples, which are usually selected by such methods as sampling from lists of inhabitants in a village, or along an accessible transect, and are usually limited to a single or small number of locales. In this study, the same spatial sample is being used for social and biological research around the park and could also be used for a range of other types of interdisciplinary work. Although this research began in 2004 and is not yet a long-term project, we believe this superpixel spatial sample has the potential to be the basis for long-term interdisciplinary research.

PARKS AND DEMOGRAPHY IN FOREST AND SAVANNA AREAS

Most of the well-known parks in sub-Saharan Africa are located in savanna environments. The problems and challenges of these parks and their surrounding populations have been relatively well documented (Brockington, 2002; Western, 1994; Newmark *et al.*, 1994; Child, 2004; Gadd, 2005). Less widely examined in the academic literature as well as less visited by tourists are forest parks (Struhsaker *et al.*, 2005). From a tourism standpoint, forest wildlife is more difficult to see, and the African forest parks that attract significant tourism are usually those with gorilla (*Gorilla gorilla beringei*) or chimpanzee (*Pan troglodytes*) populations. Even when these areas are relatively well known, the number of visitors they can accommodate is often very constrained.

The human landscapes around these parks, however, are often the reverse of their "tourist demography." Whereas human populations around savanna parks usually are limited by low and sporadic rainfall, which strongly constrains crop agriculture, forest parks in the tropics, particularly those within a broad range of mid-altitudes, often occupy and/or are surrounded by land that is highly suitable for agriculture. As a result, even when originally established (often as Forest Reserves) in a region of relatively low human population, these areas tend both to attract agricultural migrants and to support considerable population growth in place.

This has probably never been as true as during the last three decades of the twentieth century, when unprecedented rates of population increase in sub-Saharan Africa led to more than a doubling of African population (approximately a 130% increase from 1970 to 2000 (United Nations, 2006)), most of whom are still in rural areas and agricultural occupations. Uganda has experienced particularly rapid population growth. The country's population almost tripled between 1970, when it had about 9.7 million people, to 2005 when it had an estimated 27.3 million people (US Census Bureau, 2006). Uganda currently is estimated to have one of the highest growth rates even among African countries, with a natural increase rate of 3.5%, a fertility rate of 6.7, over 60% of its population under the age of 20, and about 80% rural population in 2005 (US Census Bureau, 2006). Each of these is among the highest levels in the world.

When Kibale Forest Reserve was originally established in 1932, it was located in an area with a fairly sparse population of mostly Batoro farmers and cattle raisers (Struhsaker,1997; UWA, 2004). In the 1950s, Bakiga farmers from densely populated southwest Uganda (mainly today's Kabale and Kisoro Districts near the Rwandan border) started to migrate to the area, particularly to the eastern side of the park (present-day Kahunge and surrounding subcounties). By the 1970s, the rate of natural population increase in Uganda and most of Africa began to accelerate significantly. Population growth from both natural increase and migration was particularly rapid during the 1970s and 1980s, and most of the landscape around KNP is now very heavily populated, mostly by Batoro and Bakiga farmers. In addition, tea estates on the western side of the park, which were neglected during much of the Amin (1971–1979) and second Obote (1980–1985) periods, were revived and expanded in the 1990s. As they revived, migrant workers have been attracted to the region, further enhancing population growth around KNP.

Although the rate of net in-migration has slowed in recent years, the result of the dual demographic processes of natural population growth and migration has been that the landscape around KNP (which was elevated to National Park status in 1993) is now densely populated. The recent 2002 census indicated that almost 240 000 people live in seven subcounties that border the park (UBOS, 2005). The park itself has become a virtual ecological island, and a satellite image shows the forest region of the park entirely surrounded by an intensive agricultural landscape. By our estimates, population density in our survey areas is about 260 people per sq. km on the western side of the park (which includes tea estate areas) and about 335 per sq. km on the eastern side. Both are very high densities, and they indicate intensive pressure on land, wood, water, and other resources.

The islandization of the park is partially mitigated by two elements: (1) the former game corridor that is now part of the park, intended to link KNP to Queen Elizabeth Park to the southwest (Fig. 12.1); (2) a network of bottomland forests and wetlands that radiates out from the park through the surrounding agricultural landscape. The former is relatively recent and its salience may increase in the future, while the latter is considerably older, but it has been diminishing rapidly in most areas. The corridor is still not evident on satellite imagery, and it will not be discussed in this chapter, though a recent study on it is available (Nampindo and Plumptre, 2005). The forest/wetland network and some of its positive and negative roles have been examined in this project and are discussed briefly below.

DOMESTICATED AND UNDOMESTICATED LANDSCAPES AROUND KIBALE NATIONAL PARK

As suggested above, the landscapes around KNP include several categories and types of land cover/land use. The two main categories we distinguish are the domesticated and relatively undomesticated components of the landscape. The domesticated component includes the various kinds of agricultural and related land, which are managed for economic and other purposes. For analysis of land cover and land use change, we distinguish four main types of domesticated land cover/land use: mixed crops, tea, pasture, and fallow land (the last two are distinguishable on the ground but not necessarily in satellite images). In general, these have increased from the mid 1980s to the present (based on analysis of satellite images that date from 1984 to 2003). Land covered by mixed crops and tea has increased from about 30% to over 40% of the area of the survey regions over this period.

The relatively undomesticated land cover category includes the park itself (which has grassland and wetland areas in addition to closed canopy forest) and a network of bottomland forests and (mostly papyrus (*Cyperus papyrus* L.) dominated) wetlands outside the park. Although relatively "undomesticated" in contrast to agricultural land, both the park and the surrounding forests and wetlands are nonetheless used and managed in various ways by the people around them – the park mainly today for wildlife preservation and tourism, and the forests and wetlands mainly for resource extraction (wood, water, and other products) as well as for tourism and wildlife conservation in a few areas. As discussed below, they represent both resources and sources of hazard to the people living near them. Analysis of satellite images indicates a substantial decline in the amount of forests and wetlands outside the park since the mid 1980s. In contrast, the forest canopy of the park has remained intact, and even increased in extent over the period (J. Southworth, M. Binford, and J. Hartter, personal communication).

IMPACTS AND RESPONSES TO KIBALE NATIONAL PARK

Parks in Africa and elsewhere confront a wide range of demographic, institutional, and other dynamic conditions in the human landscapes that surround them. Successful conservation that can be sustained over time is likely to require an understanding of the multiple impacts of, and responses to, parks by people living around them, which will vary from one context to another and change over time. The following findings from our survey research illustrate some of these multiple impacts and responses in the specific case of Kibale National Park. The very high population around KNP helps to foster both positive and negative interactions between the park and the surrounding communities. It also can facilitate or obstruct various conservation policies and efforts both within and outside the park.

Hazards

As elsewhere in the world, farmers and others who live near parks, or other habitat harboring wild animals, often face a variety of risks from those animals, including crop raids, livestock loss, and even human fatalities (de Boer and Baquete, 1998; Gillingham and Lee, 1999). In the case of the large number of people living around KNP, several of the primates as well as elephants (*Loxodonta africana)*, and occasionally birds can endanger people's crops and sometimes their domestic animals (Naughton-Treves, 1997, 1998; Chiyo *et al.*, 2005). In some cases, they or their children may also be threatened. People living around KNP cite elephants, baboons (*Papio anubis*), and several smaller monkeys (particularly redtail (*Cercopithecus ascanius*) and vervet monkeys (*Cercopithecus aethiops*)) as their main crop pests (Hill, 1997; Hartter, 2007). Although they appear much less frequently than primates, and over a more restricted range, elephants can almost totally destroy a field when they appear. Baboons and smaller monkeys also cause significant losses, and they can appear in substantial numbers and over a larger territorial range than elephants (particularly the smaller monkeys). Since the large majority of people in the area are farmers, these and other crop pests can be major threats to their livelihoods. Compensation for crop losses to animals is not permitted under Ugandan law – and indeed given the number of people around KNP and other parks, it seems unlikely to be practical in the future.

Almost 75% of our sample of 130 respondents said they have problems with crop-raiding. Most also said their fields must be guarded for extended periods when crops are nearing harvest. Either family members (including children) or hired workers guard fields to try to chase away animal crop-raiders. In both cases, the household incurs costs – either the labor or school time lost when a family member must guard the fields or the payments to hired guards. Most farmers report that some losses occur even when fields are guarded, either because they cannot protect everything or because they cannot chase animals away when they do appear. In most cases, killing animals is prohibited, with the exception of those, such as small monkeys and baboons, which are considered vermin (and even then, the culling of these vermin is governed by community bylaws).

Among our other findings was that both the type and extent of crop-raiding is significantly differentiated by location. Many previous studies of crop-raiding in the area were limited to the immediate perimeter of the park (Naughton-Treves, 1996). Using a larger 5 kilometer zone from the park, as well as examining how crop-raiding fits within a broader risk profile, has resulted in a more complete picture of the nature and role of crop-raiding hazards among those living around KNP.

The animals that raid farms in the area live both in the park itself and in the network of small bottomland forests and wetlands outside the park. People and farms that are relatively distant from either often have few or no crop-raiding problems at present. Farms that are buffered from the nearest wetlands and forests by two or three neighboring farms also have fewer or no incidences of crop-raiding. Elephants pose a significant hazard mainly to those who live relatively near the park (< 2 km from the park boundary) or near to one of the bottomland forests that extend directly out from the park. Baboon raiding is reported as much as 3 kilometers from the park itself, but the more distant areas are also within 500 meters or less from a bottomland forest or wetland. Small monkeys are dispersed most widely through the landscape, and raids by them are reported 5 kilometers or more from the park and more than 500 meters from forest/wetland areas. As a result, small monkeys were cited far more frequently in our survey (though not in previous studies) as the most important type of crop pest than any other animal. In many cases, these small monkeys live permanently in the small forests and wetlands rather than in the park itself. This reflects the dual character of these natural areas outside the park. While they harbor animals such as these that can pose significant risks to farmers, they also provide major resources, particularly wood, water, and natural materials (papyrus, thatch grass, etc.) for a range of other uses.

Compared to other types of parks, the areas around KNP over which wild animals pose crop-raiding or related risks are not particularly large. Risks from elephants, wild pigs, and herd animals can extend for tens or even hundreds of kilometers beyond the boundaries of many savanna parks. In this case, however, the highly populated and thoroughly domesticated landscape around KNP restricts the ranges of many wild animal species, particularly the larger animals that live mainly within the park itself.

The current geographically restricted extent of animal habitat and crop-raiding risk is a sharp contrast to the situation in the past near KNP, as reported by a sample of older respondents whom we interviewed. These respondents, most of whom had been in the area for four decades or more, all reported that much greater numbers and types of wild animals were prevalent in the past, posing serious risks to crops, animals, and occasionally to people's lives. Some of the main problems of the past, such as wild pigs (as well as large carnivores), virtually have been eliminated in the area, mostly through hunting. Indeed, hunting wild animals and consuming or selling their meat was common in the past, but is fairly rare at present, both because of scarcity of some of the main marketable species and the legal prohibition on hunting. High human population and the extensive domestication of the landscape around the park have also removed suitable habitat for many wild species, other than some of those that can extract food from farms and other human spaces. On the whole, farmers see this as a positive and important change in the landscape. As noted below, this also relates to one of the benefits that people perceive the park to have provided.

While crop-raiding is important, it is only one of many hazards that farmers and others in the area face. Our surveys included evaluations of the range of risks people faced, and among our findings were that, while widely cited, crop-raiding is not the risk people consider most significant. Among 69 respondents in a hazard survey, only 13% considered crop-raiding their most important risk, and another 26% considered it to be their second most important risk. Far more consequential are the risks and impacts of illness, which over 50% of respondents considered the most important risk they face, and another 30% consider their second most important risk. Moreover, as a number of respondents noted, they can cope at least moderately effectively with crop-raiding by guarding their fields, but the many potential illnesses they and their families face leave them with far fewer options, and can cost them considerable time and wealth (for medicine and care and, if necessary, for funerals). Other risks local farmers face include rainfall (too little or too much), theft, animal disease, and insects and other crop pests and diseases, none of which are as highly ranked as family illness or crop-raiding.

Findings such as these indicate the importance of the risks that both a park and "natural areas" outside a park (which extend well beyond the boundaries of the park itself) pose to people living near them. These can be important for both scientific and policy objectives, and they need to be taken into account in trying to enlist local people's support for conservation of such areas. In addition, a broader examination of the range of risks and other problems faced by people around a park can suggest a correspondingly broader range of potential policy measures that can be taken to compensate local people for the negative effects that the continued presence of the park and other natural areas may hold for them. The following sections discuss some current compensation efforts as well as attitudes to the park that people in our survey expressed.

Revenue sharing and "community conservation"

Since 1996, the Ugandan Wildlife Statute has mandated that a proportion of revenues collected by parks be shared with local communities (Archabald and Naughton-Treves, 2001). At present, 20% of entrance fees paid by visitors to KNP are made available to communities near the park. (KNP also collects fees for chimpanzee tracking, which are not shared with communities.) These are disbursed on the basis of proposals submitted by community leadership groups. Funds have often been used for improving community infrastructure, including schools and other buildings, among other purposes. This has been a progressive step in extending some of the financial benefits of parks to local communities, and the availability of these funds seems to be appreciated by local governance and leadership groups. However, fewer than 10% of our survey respondents cited these kinds of community infrastructure improvements as a direct benefit of the park (see below), and even fewer explicitly mentioned that these result from funds that are being returned to local communities. This suggests, at the least, that information about community revenue sharing and the use of these park earnings may not be disseminated widely.

Moreover, unlike the southern African community conservation model, there is little or no broad-based community control or decision making about the use of these funds. Perhaps even more important, the total amounts involved are very small, especially in relation to the number of people living near the park. While there has been a steady increase in the number of visitors to KNP, it currently attracts slightly fewer than 6000 visitors per year, according to park data. There are no lucrative hunting licenses available as in many southern African cases, and the tracking fees, which exceed gate receipts, are not shared with communities. Moreover, there are far more people living near KNP – almost a quarter million (UBOS, 2005) – than are found near much larger protected areas in southern Africa. As a result, it would seem that, while some of the park revenue could be channeled to local needs, the park has limited potential to serve as a major engine for local development.

Attitudes to the park

It would be understandable, given the limited benefits and relatively widespread costs, restrictions, and risks posed by the park, that those living near it would view the park in hostile or at least highly negative terms. Our surveys included several assessments of people's attitudes to the park, and these yielded some surprising results. A sample of 130 respondents was asked whether they felt the park overall helped and/or harmed them and their households. (Responses could include that the park had both helpful and harmful effects.)

Over 60% of respondents said that the park helped their household, while only 34% said the park harmed their household. Both the high number of positive assessments and the relatively low number of negative assessments were unexpected. Predictably, the highest proportion of respondents who reported that it was harmful live close to the park boundary. About 20% of respondents said the park had neither negative nor positive effects on their household.

Respondents were also asked in what ways the park has helped and/or harmed them (more than one response was permitted). The predominant negative effect, as expected, was crop-raiding, which was cited by almost 90% of those who said the park had harmful impacts. Restricted access to resources, the second most widely cited negative effect, was mentioned by only about 10% of respondents, while land expulsions and increased land costs were each cited by only about 2% of respondents.

The reasons the park was perceived as helpful were more surprising. Most literature on park and conservation policy emphasizes the importance of economic returns or rewards: employment, direct payments to communities, infrastructure development, and other related effects. However, the very high population surrounding Kibale National Park, combined with the relatively limited income and employment the park generates, as well as the small proportion of park income that can be returned to local communities (particularly to those who bear the greatest crop-raiding risks), make it very difficult for KNP to generate a significant financial or material return to local people.

This was reflected in the survey responses. Only 15% or fewer respondents mentioned any of the possible material benefits: employment of local people, improved infrastructure or educational facilities, or resources available from the park. Rather than material benefits, the main positive impacts cited were aspects of environmental services that respondents felt resulted from the presence of the park. About 60% of respondents said they believed the park brings rain to the region, and about 20% believed the park improved fresh air or the climate in general. Interestingly, Solomon (2007) found similar perceptions of the park's positive climatic and environmental impacts in her research near the southern tip of the park (currently in the corridor to Queen Elizabeth National

Park) – an area quite distant from, and with a different ethnic and environmental character from, our survey regions. In addition, over 40% of respondents mentioned that the park has helped "keep wild animals" within its boundaries. They often explained the latter by noting that wild animals were scattered throughout the landscape before the park was established, but since then animals have been "confined" within the park. While this is a questionable interpretation of the ecological history of the region, it reflects the decline of animal populations in the increasingly domesticated landscape around the park, and it represents a kind of "ethnoecology" that affects people's outlooks on the role of parks.

Little, if any, of the literature on parks reflects these kinds of local perceptions. Among other things, they illustrate that people's attitudes to parks may be shaped not solely by direct material benefits or costs, but also by a sense of broader, more diffusely distributed environmental and other impacts. Environmental education may, in this regard, have more significant and widely dispersed impacts than commonly believed, and the long-term presence of a research station such as the Makerere University Biological Field Station in KNP can help provide some of the expertise as well as data that can be the basis for ongoing environmental education in the region. While similar results might not be replicated in areas where parks have had more controversial and geographically more widespread effects, they demonstrate the importance of empirically examining the range of local perceptions, which are likely to vary from one context to another and may include some unexpected features.

CONCLUSIONS: DEMOGRAPHY AND CONSERVATION

The Kibale case illustrates, among other factors, the role of demography in the landscape around the park in determining some of the conservation potential and constraints of the park itself. The very high population around the park has meant, for example, that the income that can be earned by the park is very limited in per capita terms and cannot be expected to serve as the basis for significant economic growth centered on the park. To the degree that such income is the basis for the kind of community controlled and funded conservation found in some parts of southern Africa, this seems an unlikely outcome in cases such as KNP.

On the other hand, the densely populated landscape around KNP also means that it is unlikely that the park will expand its boundaries significantly. This helps foster a sense of stability among people living around the park, and removes one of the main reasons that parks cause anxiety and resentment in some areas of Africa (McCabe, 2002; Davis,

2006). On the whole, attitudes to Kibale National Park among the people living around it seem to be surprisingly positive. In addition, the main benefits of the park people cite are environmental services rather than economic or material benefits. Among other things, this illustrates the potential importance of environmental education among the people and communities living near parks.

Rather than on the park itself, the greatest resource pressure of the high population around the park has been on the undomesticated portions of the landscape: mainly the network of bottomland forests and wetlands that surround the park. They are likely to continue to diminish both in size and ecological integrity as population and resource pressures continue to grow, and they are, and probably will continue to be, replaced by an even more domesticated landscape. This will have important effects on the animals as well as the plants and the hydrology of these regions in the future.

The risk survey shows that crop-raiding risks are widespread, and they include risks from animals both within the park and in the wetlands and small forests outside the park. However, these are not necessarily the most important risks and problems people face. The predominance of health-related risks expressed by our respondents suggests that, if governments or conservationists are interested in assisting people living around parks, and compensating them for the costs and risks they bear in these areas, then addressing health problems might be an effective way to do so. It may not be possible or feasible to eliminate crop-raiding risks near parks, but creating zones of enhanced health services in these areas might be an effective way of generating positive support for parks and conservation among local communities.

Finally, our research has revealed some of the considerable diversity that is found around the park – diversity in land use intensity and technique, in impacts of and attitudes towards the park, in the presence of undomesticated animals and plants in the surrounding landscape, and in resource use and needs based on these. The random spatial sampling technique we have devised has proven very suitable for multidisciplinary research in this kind of highly populated region, and can be the basis for continuing long-term social as well as ecological research. Such research has the potential both to enhance conservation and to help improve people's lives.

SUMMARY

This chapter is based on some of the results of a series of surveys of local residents done in two areas around Kibale National Park. The surveys used a novel spatial sampling technique that is well suited to long-term interdisciplinary research, especially in highly populated areas such as this. Among other topics, the research examined the impacts of, and responses to, the park by people living within 5 km of the park boundaries. It showed that crop-raiding by wild animals is a widespread hazard through most of the area, though much of the raiding is conducted by animals that live in the wetlands and small forests outside the park rather than just by animals that live in the park itself. Those areas are also sources of wood, water, and other materials for local people, but extraction and/or conversion of these areas is leading to their rapid decline. Although many respondents noted the hazards of crop-raiding, illness is seen as a far more serious hazard by the large majority of people. This suggests that some compensation for crop losses might be accomplished through improvement of health care services in the area. Sharing of park revenue with local communities is an important recent innovation, but only a small proportion of respondents were aware of this or considered it a significant benefit of the park. When asked about the park's overall impacts, a surprisingly small proportion of respondents felt the park had predominantly negative impacts on the area, and a surprisingly large majority felt that the park overall had beneficial effects. Equally surprising was that the main positive impacts cited were environmental services rather than material or financial benefits. This probably reflects the low financial returns of the park relative to the very large population of the area as well as the unexpected degree to which people value widely perceived but generalized environmental benefits of forests and other undomesticated areas. Negative impacts that have been widely discussed in social science literature, such as expulsions and resource access restrictions, were only noted minimally by respondents. This suggests that education programs focusing on the general environmental benefits of forests and other natural areas can help enlist support for forest conservation, especially in highly populated areas where forested land has become scarce, and material and financial benefits or costs of conservation are not the sole, and may not be the main, determinants of people's attitudes to parks and conservation.

ACKNOWLEDGMENTS

The authors gratefully acknowledge the financial support of the US National Science Foundation, Award number 0352008, which made the research for this chapter possible. Additional support was provided by the College of Liberal Arts and Sciences and the School for Natural Resources and Environment at the University of Florida. Valuable collaboration and assistance were provided by Colin and Lauren Chapman, Patrick Omeja, Dennis Twinomugisha, and Aventino Kasangaki, as well as by our field assistants, Agaba Erimosi and Mwesige Peace. In addition, Makerere University Biological Field Station, the Uganda Wildlife Authority, and the Uganda Council for Science and Technology, and many local officials, provided useful assistance and granted permission for the research. Finally, we wish to thank the many farmers who were willing to tolerate our questions and tell us a bit about their lives.

REFERENCES

Archabald, K. and Naughton-Treves, L. (2001). Tourism revenue-sharing around national parks in Western Uganda: early efforts to identify and reward local communities. *Environmental Conservation*, **28**, 135–149.

Brockington, D. (2002). *Fortress Conservation: The Preservation of the Mkomazi Game Reserve, Tanzania*. Bloomington: Indiana University Press.

Child, B. (2004). *Parks in Transition: Biodiversity, Rural Development and the Bottom Line*. London: Earthscan.

Chiyo, P. I., Cochrane, E. P., Naughton, L., and Basuta, G. I. (2005). Temporal patterns of crop raiding by elephants: a response to changes in forage quality or crop availability? *African Journal of Ecology*, **3**, 48–55.

Davis, A. (2006). Personal communication (based on Ph.D. research around Tarangire National Park, Tanzania).

de Boer, W. and Baquete, D. (1998). Natural resource use, crop damage and attitudes of rural people in the vicinity of the Maputo Elephant Reserve, Mozambique. *Environmental Conservation*, **25**, 208–218.

Fabricius, C. and Koch, E. (2004). *Rights, Resources and Rural Development: Community-based Natural Resource Management in Southern Africa*. London: Earthscan.

Gadd, M.E. (2005). Conservation outside parks: attitudes of local people in Laikipia, Kenya. *Environmental Conservation*, **32**, 50–63.

Gibson, C. C. (1999). *Politicians and Poachers: The Political Economy of Wildlife Policy in Africa*. Cambridge, UK: Cambridge University Press.

Gillingham, S. and Lee, P. (1999). The impact of wildlife-related benefits on the conservation attitudes of local people around the Selous Game Reserve, Tanzania. *Environmental Conservation*, **26**, 218–228.

Guyer, J. and Richards, P. (1996). The invention of biodiversity: social perspectives on the management of biological variety in Africa. *Africa*, **66**, 1–13.

Hartter, J. (2007). Landscape change around Kibale National Park, Uganda: impacts on land cover, land use, and livelihoods. Unpublished Ph.D. dissertation. University of Florida, Gainesville, FL.

Hill, C. (1997). Crop-raiding by wild vertebrates: the farmer's perspective in an agricultural community in western Uganda. *International Journal of Pest Management*, **43**, 77–84.

Hulme, D. and Murphree, M. (2001). *African Wildlife and Livelihoods: The Promise and Performance of Community Conservation*. Oxford: James Curry.

McCabe, J. T. (2002). Giving conservation a human face? Lessons from 40 years of combining conservation and development in the Ngorongoro Conservation

Area, Tanzania. In *Conservation and Mobile Indigenous Peoples: Displacement, Forced Settlement and Conservation*, ed. D. Chatty and M. Colchester. New York, Oxford: Berghahn Books, pp. 61–76.

Nampindo, S. and Plumptre, A. (2005). A socioeconomic assessment of community livelihoods in areas adjacent to corridors linking Queen Elizabeth National Park to other Protected Areas in Western Uganda. Wildlife Conservation Society.

Naughton-Treves, L. (1996). Uneasy neighbors: wildlife and farmers around Kibale National Park, Uganda. Ph.D. dissertation, University of Florida, Gainesville, FL.

Naughton-Treves, L. (1997). Farming the forest edge: vulnerable places and people around Kibale National Park, Uganda. *Geographical Review*, **87**, 27–46.

Naughton-Treves, L. (1998). Predicting patterns of crop damage by wildlife around Kibale National Park, Uganda. *Conservation Biology*, **12**, 156–168.

Neumann, R. (1998). *Imposing Wilderness: Struggles over Livelihoods and Nature Preservation in Africa*. Berkeley, CA: University of California Press.

Neumann, R. (2001). Africa's 'last wilderness:' reordering space for political and economic control in colonial Tanzania. *Africa*, **71**, 641–665.

Newmark, W., Manyanza, D., Gamassa, D.-G., and Sariko, H. (1994). The conflict between wildlife and local people living adjacent to protected areas in Tanzania: human density as a predictor. *Conservation Biology*, **8**, 249–255.

Solomon, J. N. (2007). People and Protected Areas: an analysis of collaborative resource management and illegal resource use in a Ugandan National Park. Ph.D. dissertation, University of Florida, Gainesville, FL.

Struhsaker, T. T. (1997). *Ecology of an African Rain Forest: Logging in Kibale and the Conflict between Conservation and Exploitation*. Gainesville, FL: University Press of Florida.

Struhsaker, T. T., Struhsaker, P., and Siex, K. (2005). Conserving Africa's rain forests: problems in protected areas and possible solutions. *Biological Conservation*, **123**, 45–54.

Uganda Bureau of Statistics (UBOS). (2005). *The 2002 Uganda Population and Housing Census, Main Report*. Kampala: Uganda Bureau of Statistics.

Uganda Wildlife Authority (UWA). (2004). *Kibale National Park General Management Plan 2003–2013*. Kampala: Uganda Wildlife Authority.

United Nations, Population Division. (2006). World population prospects: the 2006 revision and world urbanization prospects. http://esa.un.org/unpp.

US Census Bureau. (2006). Uganda. International Data Base (IDB). Population Division/International Programs Center. Washington DC, USA. http://www.census.gov/ipc/www/idbnew.html.

Western, D. (1994). Ecosystem conservation and rural development: the case of Amboseli. In *Natural Connections. Perspectives in Community-based Conservation*, ed. D. Western and R. W. Wright. Washington DC: Island Press.

ENDNOTE

[1] We use the term "owner" to refer to a claim of permanent ownership, recognized by community bylaws, but the owner may or may not have official legal title to this land.

FRED BABWETEERA, VERNON REYNOLDS, AND
KLAUS ZUBERBÜHLER

13

Conservation and research in the Budongo Forest Reserve, Masindi District, Western Uganda

INTRODUCTION

This chapter is divided into three sections. In the first section we describe the history of conservation efforts from its beginnings as the Budongo Forest Project (BFP) up to approximately the year 2000. In the second section we describe more recent and ongoing conservation work by the Budongo Conservation Field Station (BCFS). In the third section we describe ongoing research at BCFS and how this impacts on conservation.

HISTORY OF CONSERVATION EFFORTS AT BUDONGO

Vernon and Frankie Reynolds began research and conservation of the chimpanzees of the Budongo Forest in 1962, when they spent 10 months in the forest making the first ever study of forest-living chimpanzees (Reynolds, 1965; Reynolds and Reynolds, 1965).

Our main finding in that early study was what later became known as the fission–fusion social system of chimpanzees. We did not focus on conservation issues in that first study, being mainly concerned with finding out what we could about the life of wild forest-living chimpanzees. During the Amin period, we did not return to Uganda.

The issue of chimpanzee conservation came to the fore abruptly in 1988, when VR was made aware by Shirley McGreal of IPPL of the fact that chimpanzees from Budongo Forest were being poached and sold. *The New Vision* of Monday October 17, 1988 featured an article on its front page, headlined "Chimp racket blown." The article was about the confiscation at Entebbe airport of two young chimpanzees bound for Dubai; it was suggested that they had come from Budongo. VR immediately started thinking about ways to protect the remaining chimpanzees in Budongo from the poachers, who were evidently active in the forest. It seemed that the best way to do this would be to set up a research project in the forest. The reasoning was that the presence of researchers in the forest would make it hard for poachers to enter the forest, kill chimpanzee mothers and then offer their offspring for sale. Ideally, the new project should be long term. This led to the founding of the Budongo Forest Project (BFP) in 1990. A detailed account of the project's achievements up to the year 2004 is to be found in Reynolds (2005).

One of the main contributions to conservation in the period 1990–2000 concerned the forest itself. Eggeling (1947) had studied the natural succession of Budongo forest, and in 1991 Andy Plumptre embarked on a detailed, 6-year study of the effects of logging on the forest and its wildlife. This work led to a number of important contributions to our understanding of the effects of selective logging on the forest's non-human primates, including the discovery that all three of the forest monkey species: *Cercopithecus mitis* (blue monkey), *Cercopithecus ascanius* (redtail monkey), and *Colobus guereza* (black-and-white colobus monkey), had benefited from the logging, which increased the extent of Mixed Forest (Plumptre and Reynolds, 1994). This was demonstrated by showing larger group sizes and higher densities of these primates in logged forest than in the unlogged Nature Reserves. Plumptre's dietary studies also showed that the black and white colobus monkeys of Budongo, unlike this species studied elsewhere, included fruits as a major component of their diet. Both these discoveries have direct implications for conservation of the forest and its primates: under the special conditions pertaining at Budongo, selective logging can increase the food supply of primates and this will be true of colobus monkeys as well as the species normally regarded as fruit-eaters.

A number of further studies with conservation implications were undertaken during the 1990s by students working with the Budongo Forest Project. Ugandan students from Makerere University played an important part in these. Fred Babweteera studied the impact of gap size caused by felling on regeneration, finding that larger gaps (caused by felling trees close to one another) led to climber tangles preventing tree regeneration. This was a vitally important finding for forest management. Other Ugandan students explored the implications of arboricide treatments on the tree structure of the forest, problems of mahogany regeneration, the effects of canopy opening on tree diversity, the ecology and diversity of small mammals in primary and disturbed forest, the long-term effects of forest management on the bird community of the forest, the intestinal parasites of chimpanzees, the importance of figs in the diet of chimpanzees, the bird communities in forest gaps, the ecology of Nahan's francolin, the diet of forest baboons, the regeneration of rattan in the forest, attitudes to ecotourism, the distribution of *Broussonettia papyrifera* (the paper mulberry tree), land-use changes around Budongo, the potential usefulness of less-used tree species in Budongo Forest, the problem of snaring in the forest, and a number of other topics all of which had conservation implications. During this time non-Ugandan students also contributed much to our understanding of Budongo Forest and its wildlife, with studies including all of the forest monkey species: *Cercopithecus mitis* (blue monkey), *Cercopithecus ascanius* (redtail monkey), *Colobus guereza* (black-and-white colobus monkey), and *Papio anubis* (forest baboon), many studies of the Sonso community of chimpanzees, and studies of the canopy arthropods, forest rodents, amphibians and the local human population around the forest. While not all of the studies by Ugandans and non-Ugandans were directed specifically at conservation issues, their combined impact gives Budongo Forest an unprecedented body of data on which conservation policies and actions can now be built.

Of prime importance in any long-term project trying to do research and conservation is the local community. Ugandan and non-Ugandan students have made a number of studies of the human population, including the use made of forest products, the demography of local villages, the nature of the local cash economy, attitudes to chimpanzees, the problem of crop-raiding by forest animals and the lack of food security among farmers living near the forest edge, use of snares and traps to catch wild animals, people's poor understanding of the law as it relates to hunting, and local attitudes to tourism.

Disease is perhaps the greatest danger facing any chimpanzee community that comes into proximity with human beings. We at our field station insist on the strictest adherence to our rules of hygiene as a safeguard against infection. We do not accept tourists at our field site, because of the danger of disease transmission. There are two tourist sites in Budongo Forest Reserve, one at Busingiro and the other at Kaniyo-Pabidi, and we encourage tourists to visit these. There remains, however, a constant danger of disease transmission to the Budongo chimpanzees from hunters setting snares and from illegal pitsawyers who are widespread in the forest.

A further important contribution to the conservation of Budongo's primates is periodic censusing. We have obtained accurate figures for the chimpanzee population of the whole of Budongo Forest Reserve (Plumptre and Reynolds, 1996, 1997) as well as for the populations of blue, redtail and colobus monkeys in the forest. These censuses have been and continue to be repeated at regular intervals in order to ascertain whether these primate populations are stable, increasing, or decreasing.

During the 1990s and early into the twenty-first century, the illegal harvesting of mahoganies has continued to deplete the forest of these fine trees (one species of *Khaya* and three of *Entandrophragma*). This has reduced the commercial value of Budongo, which makes the forest of less interest to the National Forest Authority because, unlike in former years, the remaining stock of mahoganies is no longer commercially viable.

RECENT CONSERVATION WORK

In acknowledgment of the conservation work of BFP, it is today an autonomous Ugandan NGO, and its name has been changed to the Budongo Conservation Field Station (BCFS). BCFS continues to blend research and conservation activities aimed at ensuring sustainable utilization of Budongo Forest Reserve. Unlike many large forest blocks in Uganda, Budongo Forest is still managed for both consumptive and non-consumptive uses. Consequently, the steps taken to ensure conservation of Budongo Forest are unique in many ways. This section highlights three current conservation initiatives: snare removal, sustainable vermin control, and forest resource harvesting regimes. All three have been born as a result of our research findings.

Snare removal

Snaring is a key threat to chimpanzee populations in many Protected Areas. Although chimpanzees are not the target for snare-setting hunters, they often get caught by this indiscriminate hunting method. In the Sonso chimpanzee community, 16/78 (21%) of the chimpanzees have snare injuries. Recognizing the magnitude of this threat, a snare-removal programme was started in January 2000. At its inception, the snare patrols covered the core range of the Sonso chimpanzee community (a total of 3896 ha) representing 9% of the total forested area of Budongo Forest. Currently, the patrols cover a total area of 13 863 ha, which represents 32% of the forested area (Fig. 13.1). The expansion followed

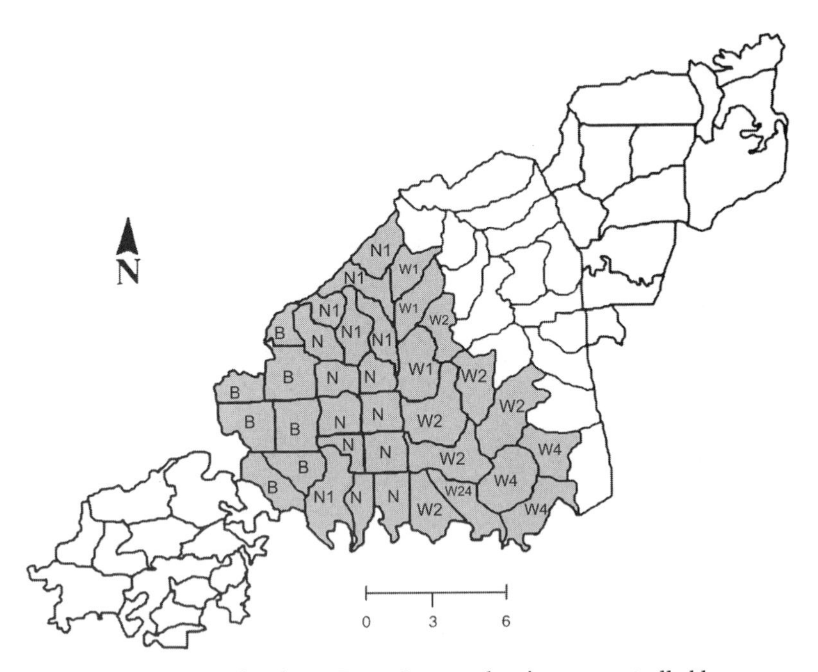

Fig. 13.1. Map of Budongo Forest Reserve showing area patrolled by snare team in 2007.

a realization that many hunters were opting to set their snares outside the patroled compartments. The average recovery rate using a four-man patrol team is 12 snares per day. The recovery rate also shows that compartments frequently used by our research team (N3, N4, N5 and N15) had the least number of snares encountered per day (Table 13.1). This further emphasizes that the presence of researchers plays a key role in deterring illegal activities in tropical forests.

To augment the efforts of the snare removal program, a community conservation education program was started in 2001 targeting forest edge communities. The aim of the education program was primarily to highlight the dangers of bushmeat hunting. However, over time, the program has encompassed other objectives such as promoting an understanding of conservation and the value of sustainable development; promoting knowledge and awareness of biodiversity and the importance of the forest and wildlife; and increasing general interest in conservation activities among the local people. Currently, the program covers 32 villages, 14 primary schools, and 3 secondary schools. In these communities, we are targeting different forest user groups as entry points to the education program. For instance, individuals involved in hunting are often

Table 13.1. *Average number of snares recovered per day in different compartments of the Budongo Forest Reserve*

Nyakafunjo block compartments		Waibira block compartments		Biiso block compartments	
Compartment	Snares/day	Compartment	Snares/day	Compartment	Snares/day
N1	11	W19	55	B1	5
N2	8	W21	6	B2	9
N3	3	W22	6	B3	13
N4	5	W23	35	B4	13
N5	3	W24	24	B5	13
N6	9	W17	62		
N7	17				
N8	16				
N15	3				

a minority among the communities. Consequently, identifying hunters as a forest user group has provided us with an opportunity to actively dissuade them from hunting by providing alternative farm activities such as pig keeping.

Sustainable vermin control

Crop-raiding by vertebrates including primates is the biggest source of conflict between rural communities and Protected Area management authorities in Uganda. Indeed, human–wildlife conflict is one of the main challenges to conservation efforts across the African continent. The human–wildlife conflict situation is particularly acute in areas of high and/or increasing population density, which is the prevalent situation around Budongo Forest. Previous research shows that primates (mainly baboons) are the most frequent crop raiders (Hill, 1997). Whenever wildlife resident in Protected Areas raid and destroy crops, any positive attitude the resident community members have towards conservation is compromised. This results in a lack of support for the protection of these areas from the local communities, jeopardizes conservation efforts, and reduces productivity of the agricultural lands. A sustainable vermin control program was initiated in 2006 with the goal of reducing incidences of human–wildlife conflicts through facilitating the identification and establishment of effective problem animal control systems. The program so far has initiated village communal guarding through the formation of village vermin control committees as well as through promotion of cultivation of crops such as chillies that are unpalatable to crop-raiders.

Forest harvesting regimes

To the local and distant communities, Budongo Forest is a major source of timber and non-timber forest products. As early as 1920, the management objectives of Budongo Forest primarily were to ensure timber production. To date Budongo is divided into three zones, namely strict nature reserve, buffer zone, and production zone. In the production zone, selective harvesting of timber and non-timber forest products is permitted. The research we have conducted to assess the effects of timber harvesting on forest regeneration dynamics primarily has benefited the forest managers (National Forestry Authority) in ensuring effective regeneration after logging. For instance, harvesting the majority of large mahogany (*Khaya anthotheca*) trees was the cause of poor regeneration of *Khaya* due to absence of seed trees (Mwima *et al.*, 2001). Moreover, maximum seed production of *Khaya* is attained at a dbh of 80 cm (Plumptre, 1995). Consequently, a recommendation was made to the forest managers to ensure maintenance of good seed trees during timber harvesting. Another study (Babweteera *et al.*, 2000) indicated that gaps created by logging are often large due to felling of multiple trees on the same site. Such large gaps are ideal sites for establishment of climber tangles, which smother tree regeneration, implying that logged areas take longer to recover. Consequently, gaps measuring 400–500 m^2 were recommended. These recommendations have been adopted and incorporated into the forest management plan.

ONGOING AND FUTURE RESEARCH

Ongoing primate research is relevant both directly and indirectly for the conservation of the Budongo chimpanzees in the following two ways. First, we are in the midst of a hormonal study to monitor the stress response of individuals to different anthropogenic factors, and for obvious reasons this line of research is straightforwardly relevant to chimpanzee conservation. A second line of research is dealing with the chimpanzees' cognitive and communicative skills, which may require some additional explanation with regard to its relevance for conservation. The next section explains in more detail how these current research efforts are relevant for conservation efforts and decisions, either by providing important data that are directly relevant, or by generating insights into the chimpanzees' minds that are likely to alter public attitudes towards chimpanzees, our closest living relatives.

Hormonal studies

Human populations continue to grow in many parts of the world, including Uganda, and habitat loss is a direct consequence and reality for many primate populations, including the chimpanzees. One way of assessing the impact of these changes is to monitor the stress responses of individuals. Stress is an interesting variable for various theoretical reasons, but it is also important for conservation efforts because in many species elevated stress levels impact directly on the health and viability of endangered populations. Thus, one of our current projects, spearheaded by Zinta Zommers (D.Phil. student, University of Oxford) is concerned with the behavioral and environmental predictors of stress in the Budongo chimpanzees. This research is done in collaboration with Dr. Tobias Deschner at the Max-Planck Institute for Evolutionary Anthropology in Leipzig, and Dr. Fabian Leendertz at the Robert Koch Institute in Berlin.

It is now possible to study hormones, including those related to stress, by analyzing the content of feces and urine excreted by chimpanzees. This technique allows researchers to study the relationship between social and environmental events and stress. It has long been known that heightened levels of stress can lead to disease, and one particular focus of this project is to understand how anthropogenic disturbance impacts on the stress experienced by chimpanzees. The project will further explore whether suspected impacts are direct, for example, due to altered food availability, or indirect, for example, due to changes in the patterns of social interaction between group members or between neighboring groups. In summary, understanding the impact of human disturbance and environmental change on stress is likely to lead to a better understanding of the conservation needs required to secure the survival of the chimpanzees in Budongo Forest and in other parts of Africa.

Cognitive studies

A second line of research, whose importance for chimpanzee conservation is less immediately obvious, has to do with the cognitive and communicative skills in this primate species. Cognitive skills, and their underlying neural systems, have evolutionary histories just like physical traits. As a result, closely related species are typically more similar in their cognitive capacities than more distantly related ones, a persuasive reason to conduct comparative studies with primates when dealing with problems of cognitive evolution. Chimpanzees naturally play a key role in any such endeavor.

One key aspect of what it means to be human concerns our extraordinary ability to communicate about events, thoughts, and emotions. For this process, humans rely almost entirely on speech, a unique orofacial motor skill that affords rapid assembly of simple vocal units into more complex utterances, the carriers of meaning. The origin of speech and language has been singled out as one of the great riddles of modern science (Holden, 2004). Most authorities now subscribe to the view that comparative animal research is the most powerful empirical tool to solve the problem of how, when, and why humans evolved language (Hauser *et al.*, 2002).

Genetic evidence suggests that our hominid ancestors may not have possessed such sophisticated vocal skills until very recently, perhaps as little as 200 000 years ago (Enard *et al.*, 2002). How exactly hominids communicated before remains unclear, but one implication is that many of the cognitive abilities required for human linguistic communication must be significantly older than speech itself, rooted deeply in the primate lineage.

Our current research with the Sonso chimpanzees is based on this premise. It aims to describe the psychological apparatus underlying natural communication in these primates. We are in the process of investigating how chimpanzees produce and interpret vocal signals in relation to social and ecological events. We are particularly interested in phenomena such as referential utterances, intentional signaling, combinatorial meaning, inferential skills, and the ontogeny of meaning. Results of these studies are likely to lead to a better understanding of a fundamental problem in science, that is, how and why humans evolved a unique trait, language. Vocal behavior has been an extremely productive tool in assessing the cognitive capacities of monkeys (Cheney and Seyfarth, 1990). Astonishingly, however, great apes have been largely neglected in this respect, possibly due to a widespread belief that the calls of our closest living relatives cognitively are uninteresting and not the product of their otherwise remarkable intelligence. Indeed, the vocal repertoire of chimpanzees consists only of a small number of graded call types, but recent studies have revealed substantial context-specific acoustic variation within these call types (e.g., Crockford and Boesch, 2003). For example, chimpanzees produce rough grunts when finding food, and the acoustic structure of this call type varies subtly depending on the type of food encountered. In captivity, it has been shown that some call variants are linked with certain food types, effectively turning them into referential signals for specific foods (Slocombe and Zuberbühler, 2006). When hearing rough grunts of other group members, a listener can draw inferences about the corresponding food type and use this information to guide his search for food (Slocombe and Zuberbühler, 2005b). The screams given by individuals during agonistic interactions have revealed similar patterns. Like the grunts, these calls are highly graded, and acoustic variants predict the social role of the individual during a conflict (Slocombe and Zuberbühler, 2005a). Moreover, callers appear to encode the severity of the aggression, allowing nearby listeners to decide whether or not to interfere in a conflict.

Chimpanzees live in individualized fission–fusion communities and one consequence is that a caller's audience constantly changes (Goodall, 1986). One of our pilot studies suggests that victims of aggression exaggerate the nature of the attack by producing high severity screams, but only if high-ranking allies are in the audience. This kind of tactical calling has not been described before for non-human primates and deserves further scrutiny. Second, there is no direct evidence that nearby individuals actively perceive the acoustic differences between victim and aggressor screams, and that they use this information in behavioral decisions. For example, when a chimpanzee hears two group members involved in a dispute, can he draw inferences about the social roles played by the opponents and, if so, will he use this information to decide whether or not to interfere in the dispute? Another important pillar of primate social intelligence concerns the ability to monitor the relations of other group members. Male chimpanzees notoriously are obsessed with social rank, presumably because of the reproductive implications, and alliances with other males are crucial in this endeavor. Consequently, changes in other group members' social ranks can have drastic consequences, so that monitoring rank changes is important (e.g., Bergman *et al.*, 2003). For example, if a young and low-ranking male suddenly begins to prevail during conflicts, will others change their social behavior towards him in the light of his potential value as a future alliance partner?

Our general approach to these questions is to use a combination of careful observations and playback techniques to assess the cognitive abilities of our chimpanzees. We have conducted playback experiments successfully with the chimpanzees at Edinburgh Zoo and currently we are conducting a first playback study at Sonso. Our experiences so far point to the fact that the technique is well suited to study the vocal behavior and cognition in free-ranging chimpanzees and that results are likely to produce progress in better understanding the mind of this species.

In recent years questions concerning the evolution of language have attracted significant interest from the scientific community as well as from the general public. The comparative approach is regarded widely as the most compelling empirical tool, and research with chimpanzees naturally plays a key role. Our research with the Budongo chimpanzees aims to address some points by targeting current theories of animal communication and language evolution at a time when there is large-scale interest in this field. Most natural populations of chimpanzees are under severe threat caused by habitat loss, diseases, and other anthropogenic disturbances. It is our hope that, by studying the minds of our closest living relatives in the wild and by revealing their complexities and sophistication, we will be able to raise awareness of these disturbing developments as well as to establish the chimpanzees as a living link to our own evolutionary past. In doing so, we hope to achieve a change of general attitude towards these primates in the general public and its policy makers, particularly in chimpanzee habitat countries. In summary, we hope that some of our results will be pivotal for raising public awareness for urgent conservation measures.

CONCLUSION

This chapter has shown how, from the start in 1990, the work of Budongo Conservation Field Station has been committed to research and conservation of the forest and its wildlife, especially its chimpanzees. From the first decade during which research was the priority, many, indeed most of the projects we undertook were relevant to conservation and provided information of value to decision makers in the forestry and environment sectors. During the second decade (from 2000 onwards) we have made a number of conservation-oriented studies of the forest and the local population around Budongo, and our applied conservation wing is now in full swing, thanks largely to the efforts of our Director, Fred Babweteera of Makerere University and our new Alfriston Centre for Conservation and Education. At the same time we have continued the research on chimpanzees that has brought BCFS its high international reputation. This research, much of it of an academic nature, is based on a number of universities, primarily the University of St. Andrews in Scotland from where Klaus Zuberbühler serves as the Scientific Director of BCFS. We hope to continue to demonstrate that research and conservation can amplify and support each other in the coming years and decades.

SUMMARY

After an initial year spent in the Budongo Forest in 1962, Vernon Reynolds founded the Budongo Forest Project (BFP) in 1990 with the help of Chris Bakuneeta and Andy Plumptre. Studies of the forest in relation to logging activities and their effect on wildlife populations characterized the work of the project in its first decade, along with more academic studies of chimpanzees, other primates and non-primate species. This work focused on comparisons of logged and unlogged forest, thereby ensuring conservation and management implications. Since 1990, the project, now a Ugandan NGO and renamed the Budongo Conservation Field Station (BCFS), has continued to include a strong conservation element in its research activities, which include studies of the local population around the forest as well as continued research on the chimpanzees and other wildlife species and their interactions with the forest. In this chapter we demonstrate how the work of BCFS has implications, direct and indirect, for conservation of Uganda's forest sector.

ACKNOWLEDGMENTS

We are indebted to a number of bodies for funding and, in particular, would like to express gratitude to the following: USAID, Jane Goodall Institute, National Geographic Society, Department for International Development, Norwegian Aid and Development, and the Royal Zoological Society of Scotland. We have also benefited from time to time from a large number of smaller agencies, and from individual donors.

REFERENCES

Babweteera, F. (2006). Interactions between frugivores and fleshy-fruited trees in primary and secondary tropical rain forests. Unpublished D.Phil. thesis, University of Oxford.

Babweteera, F., Plumptre, A. J., and Obua, J. (2000). Effect of gap size and age on climber abundance and diversity in Budongo Forest Reserve, Uganda. *African Journal of Ecology*, **38**, 230–237.

Bergman, T. J., Beehner, J. C., Cheney, D. L., and Seyfarth, R. M. (2003). Hierarchical classification by rank and kinship in baboons. *Science*, **302**, 1234–1236.

Cheney, D. L. and Seyfarth, R. M. (1990). *How Monkeys See the World: Inside the Mind of Another Species*. Chicago: Chicago University Press.

Crockford, C. and Boesch, C. (2003). Context-specific calls in wild chimpanzees, *Pan troglodytes verus*: analysis of barks. *Animal Behaviour*, **66**, 115–125.

Eggeling, W. J. (1947). Observations on the ecology of the Budongo rain forest, Uganda. *Journal of Ecology*, **34**, 20–87.

Enard, W. (2002). Intra- and interspecific variation in primate gene expression patterns. *Science*, **296**, 340–343.

Goodall, J. (1986). *The Chimpanzees of Gombe: Patterns of Behavior*. Cambridge: Harvard University Press.

Hauser, M. D., Chomsky, N., and Fitch, W. T. (2002). The faculty of language: what is it, who has it, and how did it evolve? *Science*, **298**, 1569–1579.

Hill, C. M. (1997). Crop raiding by wild vertebrates: the farmer's perspective in an agricultural community in Western Uganda. *International Journal of Pest Management*, **43**, 77–84.

Holden, C. (2004). The origin of speech. *Science*, **303**, 1316–1319.

Mwima, P. M., Obua, J., and Oryem-Origa, H. (2001). Effect of logging on the natural regeneration of *Khaya anthotheca* in Budongo Forest Reserve, Uganda. *International Forestry Review*, **3**, 131–135.

Plumptre, A. J. (1995). The importance of seed trees for the natural regeneration of selectively logged tropical forests. *Commonwealth Forestry Review*, **74**, 253–258.

Plumptre, A. J. and Reynolds, V. (1994). The effect of selective logging on the primate populations in the Budongo Forest Reserve, Uganda. *Journal of Applied Ecology*, **31**, 631–641.

Plumptre, A. J. and Reynolds, V. (1996). Censusing chimpanzees in the Budongo Forest, Uganda. *International Journal of Primatology*, **17**, 85–99.

Plumptre, A. J. and Reynolds, V. (1997). Nesting behaviour of chimpanzees: implications for censuses. *International Journal of Primatology*, **18**, 475–485.

Reynolds, V. (1965). *Budongo: A Forest and its Chimpanzees*. New York: Methuen, and London: Doubleday.

Reynolds, V. (2005). *The Chimpanzees of the Budongo Forest: Ecology, Behaviour and Conservation*. Oxford: Oxford University Press.

Reynolds, V. and Reynolds, F. (1965). Chimpanzees of the Budongo Forest. In *Primate Behavior: Field Studies of Monkeys and Apes*, ed. I. DeVore. New York: Holt, Rinehart & Winston.

Slocombe, K. E. and Zuberbühler, K. (2005a). Agonistic screams in wild chimpanzees (*Pan troglodytes schweinfurthii*) vary as a function of social role. *Journal of Comparative Psychology*, **119**, 67–77.

Slocombe, K. E. and Zuberbühler, K. (2005b). Functionally referential communication in a chimpanzee. *Current Biology*, **15**, 1779–1784.

Slocombe, K. E. and Zuberbühler, K. (2006). Food-associated calls in chimpanzees: responses to food types or food preferences? *Animal Behaviour*, **72**, 989–999.

ANTHONY COLLINS AND JANE GOODALL

14

Long-term research and conservation in Gombe National Park, Tanzania

ORIGIN AND HISTORY

Louis Leakey hoped that a study of chimpanzees, our closest living relatives, living on the shores of a lake might give clues as to the behavior of Miocene hominoids living on the shores of Lake Victoria, on the islands of Rusinga and Mfangano. Accordingly, he arranged for Jane Goodall to start observing the chimpanzees of Gombe on the eastern shore of Lake Tanganyika in July 1960.

As it turned out, the proximity of the lake was not relevant, but information about their behavior was of great significance. Like our earliest ancestors, chimpanzees *(Pan troglodytes)* hunted for meat and shared the kill. And they used and made tools, a behavior believed to be unique to our own species – so that Leakey gleefully suggested that we must redefine man, redefine tool, or include chimps as human! As a result of these and other fascinating discoveries, both Jane Goodall and the Gombe chimpanzees gained a good deal of publicity around the world (Goodall, 1965, 1971). This attracted funding from a succession of donors. Other scientists and students arrived to work at Gombe, and the Gombe Stream Research Centre (GSRC) was established in 1964. The study has continued from 1960 until the present day: there have been over 200 scientific papers, 35 Ph.D. theses, over 30 books, nine films, over 160 popular articles and secondary writings, and hundreds of lecture tours and conferences; and students trained at Gombe have moved on to study chimpanzees elsewhere, as well as different species of primates and other animals in other countries. Many of them are also involved in conservation programs.

This high profile has inspired and encouraged countless young people to enter the fields of research and conservation. In addition, at the local level, the research at Gombe has led to a greater awareness of conservation issues in the surrounding villages, and this in turn has resulted in a number of initiatives that benefit the environment, the chimpanzees and the quality of life of the villagers living near the National Park.

Conservation came out of the research in three phases: first, direct conservation benefits arose from the studies done; second, we then augmented these by outreach and collaborations from the research, both formal and informal; and third, we developed formal planned conservation programs, which are now being replicated in other areas.

PHASE 1: DIRECT CONSERVATION BENEFITS FROM THE RESEARCH

Although many of the research questions investigated at Gombe Stream Research Centre were framed without reference to conservation realities, many have produced results of value toward conservation, which are described extensively by Pusey *et al.* (2007).

Upgrading the conservation status of the study area

Gombe had been demarcated as a Game Reserve in 1943, ahead of any research, to protect the chimpanzees living there. In the 1960s the forest stretched for miles along the eastern shores of Lake Tanganyika, but from the late 1970s onward, human population growth, exacerbated by the influx of refugees from Burundi and then from Congo, led to massive deforestation in the area. Fortunately, because of the interest in the chimpanzees that resulted from the research in the early 1960s, the 30 sq. mile reserve was gazetted in 1968 as a National Park (Gombe National Park), thus protecting the area from destructive human activities. The Gombe Stream Research Centre continued as a separate institution within Gombe National Park.

Population changes, and threats to chimpanzees

The long-term project identified individuals within the main study community very early (Goodall, 1986). Other communities in the Park were not investigated until later, but by 2000 almost every individual within the Park had been identified (Greengrass, 2002–3), revealing population composition and birth rate.

First, this shows that there are currently about 100 chimpanzees. This is already a cause for concern, as previous estimates had been around

150: the major declines have occurred in the northern and southern communities, mainly from habitat loss outside the Park. A population of 100 is also below the theoretical minimum viable for conservation, once chimpanzees' slow maturation, low birth rate, and survivorship are taken into account.

Second, life histories of individuals give good indication of the main causes of reproductive failure and death, and these are disease, intergroup aggression (with small communities at especial risk), poaching, and habitat loss outside the Park boundary (Pusey *et al.*, 2007).

This list of threats is valuable because it can dictate the best strategies for conservation at Gombe. But, in addition, each threat can be assessed for its severity in other chimpanzee populations in different unprotected areas, as a starting point for their conservation.

Pusey *et al.* (2007) also emphasize that chimpanzees' intense territoriality means that captive chimpanzees cannot be returned to wild populations, since resident chimpanzees are likely to kill them. Also, chimpanzee communities need much space, between 5 and 20 sq. km at Gombe, even more elsewhere, which means that demarcation of small conservation areas such as Gombe (35 km^2) is no guarantee of success, unless measures can be taken to increase their range even outside the Park.

Knowledge of biodiversity

Several Gombe studies have identified primates' food plants, and non-foods, amounting to a comprehensive list of plant species now valuable as a biodiversity inventory. Other species lists include birds, reptiles, and termites (Bygott, 1992).

Vegetation change, and habitat loss

Use of satellite imagery to investigate forest changes within the Park also inevitably revealed the severity of deforestation outside (Pintea *et al.*, 2003).

Protection from poaching

Gombe's linear shape means the chimpanzees are particularly exposed: nowhere inside the Park is more than 1.6 km from the edge, and houses come very close to the boundary (see Fig. 3 in Pintea, 2006). However, researchers moving with the chimpanzees provide protection. Poaching was worst in the community that was studied the least (Greengrass, 2002–3).

PHASE 2: OUTREACH PROMOTING CONSERVATION

Informal outreach

The research staff at Gombe is mainly Tanzanian, employed from the villages closest to the Park. This is rare employment in an economically poor rural community, and it gives the villagers a vested interest in the survival of the chimpanzees. It also helps to build up community awareness, as the field staff become deeply involved in the chimpanzees as individuals, and share stories with their families and neighbors.

Formal outreach

Wildlife awareness weeks and school visits

Research staff have organized a series of wildlife weeks in the nearest town, and elsewhere, with several events and exhibits, featuring the chimpanzees, the threats to their survival, environmental degradation, and how to address these threats. School parties visited the exhibit and subsequently have been on many excursions to the Park. It was at such an awareness week in Dar es Salaam in 1991 that the youth program, Roots & Shoots, was created (see below).

Education outside the Park – Gombe Research Education Program (GREP)

From the 1970s, the chimpanzee community in the southern part of the Park began to decline severely, because refugees from Burundi and Eastern Congo settled nearby, clearing forest and woodlands bordering the Park, poaching, and setting bushfires (Greengrass, 2002–3). Accordingly, the Gombe Research Education Program, with UNDP funding, held a series of village meetings to discuss the situation, and gave reasons for protecting the chimpanzees and their habitat. These meetings were facilitated jointly by the Research team, the Community Conservation team (TACARE, below), and Gombe National Park's Outreach (CCS). Villagers were also invited into the Park and shown the rich fauna and flora.

World outreach and tourism

Support for conservation

Much information about chimpanzees has flowed from Gombe research into the world market through books (including bestsellers, Goodall, 1971, 1999), films (e.g., National Geographic Society from 1963, Animal Planet from 2003, an IMAX film with Science North 2002), and public lectures, a great many by Jane Goodall herself, calling attention to the endangered status of chimpanzees. These, with the many scientific papers and talks, have aroused concern for the plight of chimpanzees, and the need to better protect them.

Tourism

Media output, public lectures, and even research results, attract tourists who bring foreign exchange to the country and money to the local community (travel, hotels, tours, and souvenirs) and give them a business interest in the chimps and their habitat. Some of the local villagers even gain employment as guides, though greater benefits need to be extended to those villages close to the Park.

Tourism attracted by these media also brings considerable revenue to Tanzania National Parks, defraying their annual costs of running Gombe National Park, thus paying in part for its conservation.

Collaborations with Tanzania National Parks

Several of the research studies have influenced management of the National Park, such as recommendations for fire control, and limiting the size of tourist groups and their behavior when near chimpanzees. The Jane Goodall Institute (JGI)–Gombe research team also contributed extensively to the writing of a General Management Plan for Gombe National Park and paying for the costs (TANAPA, 2005).

Also, the Research team's conservation initiative in the south of Gombe (above) assisted the National Park with radio communication and boat transport, from UNDP funds, and the Greater Gombe Ecosystem program (see below) assists with anti-poaching measures through a USAID grant.

PHASE 3: PLANNED CONSERVATION PROGRAMS

The research within Gombe, and the researchers' accumulated knowledge of the local environment and community, have led to five separate conservation initiatives run by the Jane Goodall Institute.

Conservation within the research program at Gombe

Gombe Stream Research Centre has one Director of Conservation, a local Tanzanian with a Ph.D. in primate conservation, to monitor the chimpanzee population and other resources within the Park, and to be aware of habitat changes and human influences from outside.

Health monitoring at Gombe Stream Research Centre

Because the main identified cause of death in the Gombe chimpanzees is disease, first priority has been given to monitoring their health. Prevention is the aim, but treatment at times is justified, because some of the most severe disease outbreaks were thought to have been transmitted from humans. This disease risk is the downside of tourism.

The Health Monitoring Program for the chimpanzees began in October 2001 (Lukasik, 2002), and it is now funded and run by Lincoln Park Zoo, Chicago (Lonsdorf *et al.*, 2006). The main components are:

- Employment of a qualified Tanzanian veterinarian, who has been given additional practical training working with chimpanzees in our sanctuaries for orphans, and other captive situations.
- Routine daily monitoring of chimpanzee health, notes of any individual showing signs of illness or injury and follow-up on any signs for concern. There is monthly fecal sampling for parasites, and 3-monthly fecal sampling for SIVcpz (the chimpanzee retrovirus related to HIV), the latter in collaboration with the University of Alabama (Sharp *et al.*, 2005).

Community-centered conservation on a landscape scale: the TACARE model

Since 1960, a high human birth rate and periodic influxes of people fleeing wars in Burundi and Congo have brought a 4.8% annual growth rate (2001 census), and a population density of 45 people per km^2 to the area around Gombe. This has led to deforestation due to charcoal burning and clearing forest for agriculture, loss of soil fertility, and serious soil erosion. This has all been well documented by our research team. In order to tackle these problems, the Jane Goodall Institute initiated the Lake Tanganyika Catchment Reforestation and Education program (TACARE), with funds from the European Union, in 1994.

TACARE was a direct outgrowth from the research program. Its first strategy was to establish tree nurseries in each of the villages around Gombe, and to encourage the creation of woodlots (for firewood) in order to decrease the rate of cutting near the Park. However, the rate of deforestation continued to increase, almost doubling around Gombe, from 1991 through 2003

(Pusey *et al*., 2007). Tree planting could not keep up. In addition, the villagers were not concerned initially with conservation of the environment because of their own immediate problems of poverty (the second poorest region in Tanzania), over-population, over-farmed infertile soils, depleted water supply, and poor health. It was clear that the TACARE team needed to take these issues into account if they were to build up good relations with the villagers. When team leaders sat down to discuss these things with them, they quickly learned that the villagers' greatest problems were lack of social infrastructures, clean water, education, primary health care, and capital. These were all rated as more important than conservation of resources. So, in 1998, with new funding, a major effort was made "to address the interrelated problems of poverty, health, and sustainable land use, in order to protect the remaining forest and biodiversity of the region," and TACARE established integrated departments to address each of these areas.

In 2002, the TACARE program was extended to cover an area of 200 km^2, delivering its services to 24 adjacent villages surrounding Gombe and along the lakeshore. This program, known as the Greater Gombe Ecosystem (or GGE), is funded by USAID and the Annenberg Foundation. Of great importance is the use of GIS and satellite images that assists each village to draw up its own Land-Use Plan (Pintea, 2006). TACARE believes that this strategy will allow regeneration of upper watershed forests and woodlands so as to ensure village water supply. This will not only help to protect natural flora and fauna, including chimpanzees, but will also allow the villagers to harvest honey, edible mushrooms, and medicinal plants.

The services delivered to each village by TACARE through this program are as follows.

Forestry

Village nursery attendants grow seedlings for reforestation, but the greater emphasis is on natural regeneration, whereby Forest Monitors control burning, tree-felling, and hunting, in Village Forest Reserves and in other remaining woodlands and forests. Villagers are trained in improved bee-keeping, and traditional healers are drawn into a cooperative to share knowledge and to locate the most important forest/woodland sites for conserving medicinal plants.

Agriculture

Agriculture workers focus on reducing shifting cultivation (slash and burn) by promoting sustainable farming techniques. This means they promote cash crops (high-yield hybrid oil palm, and quality coffee linked to speciality markets in the USA). They use demonstration plots to promote the agroforestry, and they control soil erosion by contour planting with *Vetiveria* grass. Through these methods, over-used farmland can become productive again within 2 years.

Health

Water engineers (funded by UNICEF) assist with construction of water supply (wells and gravity schemes), and ventilation-improved pit latrines (VIP); and they provide sanitation and hygiene training (PHAST).

Health specialists train a network of volunteers in the villages (CBDAs) who deliver information and assistance with family planning, and, in collaboration with district medical officers, provide information and counselling on HIV, and related interventions and home-based care.

Community development

The project officer organizes the construction of classrooms, dispensaries, and other needed community buildings. Community development officers promote fuel-efficient stoves, and set up savings and credit schemes for income generating activities. HIV-positive villagers are encouraged and helped to join these schemes. This department also funds scholarships for girls that enable them to go through secondary school, including orphans and economically poor girls who consistently have performed well in their studies.

Land-use planning and GIS

The Land-use planner and GIS specialist together facilitate participatory Land-use management, by locating and mapping village boundaries using GPS, and helping the villagers to agree these boundaries has allowed to each village begin their participatory land-use plan, assisted by satellite images and GIS. Initially, this will be in the 14 villages closest to Gombe National Park.

Youth programs

In TACARE villages, young people are encouraged to involve themselves in environmental issues, including reforestation, erosion control, sustainable agriculture, responsible animal husbandry, and bushmeat reduction. They also address social issues such as HIV and teenage pregnancy, and make excursions to forest areas including Gombe. These are all provided through the Roots & Shoots program (see below).

Outcome

The Greater Gombe Ecosystem Program will favor Chimpanzees as watershed woodlands will increase the food supply, but ideally they will also become connected as corridors, which may enable chimpanzees to move between Gombe and other isolated relict populations. This is the only way that these small chimpanzee groups might be saved from the adverse effects of inbreeding.

TACARE's Community Conservation Model replicated elsewhere

The same TACARE model is currently being replicated in two other chimpanzee habitats, one in Tanzania, and another in eastern Democratic Republic of the Congo.

Masito–Ugalla Ecosystem

Masito and Ugalla are areas of forest and miombo woodland 60 km south of Kigoma, harboring numerous chimpanzees, which have not yet been subject to the same pressures from human population as at Gombe. Chimpanzees of Masito have already partially been surveyed, at Filabanga and Kasakati. The Ugalla chimpanzee population is of particular interest as the dry environment represents the limits of the species' ecological range (see refs. in Massawe, 1992).

In 2005, GSRC (S. Kamenya) initiated a community conservation project in Masito (funded by UNDP until 2008). This program, with new funding from the Pritzker Foundation, will now be expanded to include Ugalla. The Masito–Ugalla Ecosystem program will cover 6000 km^2. The TACARE model of community involvement will be used to safeguard adequate habitat for the chimpanzees, whilst improving conditions for the human population.

The TACARE model of community-centered conservation in the Democratic Republic of the Congo

In 2005 JGI moved into eastern Congo, as part of a consortium of NGOs: Conservation International (CI), the Dian Fossey Gorilla Fund

International (DFGFI) and the Union of Associations for the Conservation of Gorillas and Community Development (UGADEC). The 35 000 km^2 landscape is demarcated by Kahuzi-Biega and Maiko National Parks and the Tayna Gorilla Reserve, and includes intervening village lands. An estimated 5000 eastern lowland gorillas (*Gorilla beringei graueri*) and 15 000 chimpanzees live in this area. The aim is to persuade villages to allow part of their lands to remain as a green corridor that will allow movement of animals between the two National Parks. JGI (funded by USAID, FAO and private donors) is replicating the TACARE model and gaining the goodwill of the villagers

Youth Program: Roots & Shoots

The Jane Goodall Institute's Environmental and Humanitarian youth program, Roots & Shoots, empowers its members, from kindergarten through university, to roll up their sleeves and take action on behalf of animals, the environment, and the human community. Its main message is that every individual makes a difference every day, and that, by acting together, young people can create positive change. It was initiated in Tanzania in 1991 and is now active in more than 20 regions of Tanzania and more than 90 countries worldwide.

Roots & Shoots is not a conventional conservation program, because it exists mainly in clubs in schools. These clubs are in all the villages around Gombe National Park and throughout the Greater Gombe Ecosystem/TACARE area, and also in other schools adjacent to chimpanzee groups in Mkongoro-Kwitanga. In these schools, Roots & Shoots has a strong emphasis on forest conservation, and the importance of chimpanzees and other wildlife.

Roots & Shoots is very active in the Refugee Camp at Lugufu, 50 km to the east of Gombe, where some 76 000 Congolese are housed (as of November 2006). The refugees' rations are without meat, and they (or people hunting on their behalf) have badly depleted the wildlife, including monkeys and chimpanzees, in the nearby Lilanshimba Hills (Ogawa *et al.*, 2006). There are now 14 Roots & Shoots groups in and close to the camp, and one of their priorities is to advocate strongly against consumption of wild meat, and to promote chicken rearing for alternative protein. A recent survey of refugee children, comparing Roots & Shoots members and non-Roots & Shoots members, attests to the success of their program (Meshach *et al.*, 2007).

ADVANTAGES AND LESSONS LEARNED

Starting small and growing slowly

A large part of the success of the TACARE project was that it started small and grew slowly. This allowed the project staff time to test different approaches until finding the one that best suited the community's needs, the holistic approach. It allowed the villagers time to appreciate the real value of what TACARE could offer whilst, at the same time, they came to know and trust the TACARE staff. TACARE has thus developed as a network of close relationships that underlie the success of the project as a whole.

Dealing with the community's perceived problems

A key element in the success of TACARE was the realization that only if villagers were bought into the program and felt a sense of ownership, could it really work. Thus, much time was spent initially in listening to their concerns, and what they felt could help them most.

Relying on villagers to implement the schemes

Since the start, TACARE has always trained villagers to implement the program. For example, the family planning and women's reproductive health work is done by volunteers. These men and women from the villages are trained periodically at the Kigoma office or in village workshops, and they then operate in their own villages. This means help is continually available to villagers in a low-key and culturally sensitive way from people from their own communities whom they trust. Involving the villagers in this way ensures sustainability.

Giving knowledge, rather than equipment and technologies

Providing equipment and technologies leads to short-term benefits and the approval of villagers, but it is expensive, and lasts only as long as the equipment. By contrast, providing knowledge, and instilling a sense of ownership and accomplishment, costs less, lasts longer, and can lead to real long-term benefits. TACARE teaches simple, affordable techniques that lead to an environmentally sustainable future.

Collaborating with the local government at village level and above

Any conservation plan that works through the local community must ensure that the local government, from village up through district and regional level, is aware of what is proposed, and is involved in what is being done. Their support is vital in many ways. After all, the land and the natural resources of the communities are theirs. It is hoped that TACARE will be taken over eventually by the local government and the people. The long-term success of conservation efforts implemented now may well depend on nurturing the right relationships with all appropriate levels of government.

PROBLEMS

Floating populations

While collaboration with the local government is necessary, it may not be sufficient. This because there may be people out of touch with the village government: floating populations, refugees, itinerant poachers, or timber harvesters. In TACARE's case, some of the more severe environmental degradations were caused by illegal immigrants living in the poorer land at the margins of the villages. It was only once these people could be contacted that the program could become fully effective.

People's fears for their livelihoods and land

One of the major obstacles to a conservation program is that people often fear that it will deprive them of their livelihoods or their land. Thus, it is essential to protect and, when possible, improve the livelihoods of the people within a community in an environmentally sustainable way, and try to promote alternative livelihoods for those people who are most destructive to the environment.

IMPROVEMENTS

Emphasis on conservation from the start of the research

A number of the threats to Gombe's chimpanzees could, in hindsight, have been reduced if appropriate safeguards had been built in from the start. The decline of the southern chimpanzee community could have been prevented if research staff had been able to assist Park staff to detect the rapid deforestation, and the increased hunting caused by refugees outside the Park boundary.

Monitoring of conservation impacts

Conservation planning and funding requests cannot be effective without good data on the progress and impact of conservation activities. Precise baseline measures from the start, and regular monitoring, assist greatly the efficient use of resources and energy.

Conservation Action Plans

JGI is now working with the The Nature Conservancy of the USA to draw up integrated Conservation Action Plans for the greater Gombe and the Masito–Ugalla ecosystems. It will be necessary to assess the relative importance of different threats to chimpanzees and their habitats, and also to try to determine the natural resources, within chimpanzee habitat, needed by the human communities. This information will then dictate priorities for the action plans of the two programs. Details are available from: http://conserveonline.org/workspaces/cap.

Training

It is essential for the long-term future of conservation that nationals receive training that will provide them with the knowledge, and the qualifications, to enable them to take over the protection of the environment and wildlife in their country.

Long-term studies, such as that on the Gombe chimpanzees, provide researchers with an excellent opportunity to detect candidates who have the necessary passion and commitment, and provide them with further training and education as required.

SUMMARY

Long-term research at Gombe has contributed to conservation in three ways. First, the research produces, and has produced throughout, information valuable to conservation. Second, the research team initiated some conservation outreach to the surrounding villages. And third, increasing understanding of the threats to chimpanzees led directly to the development of a unique integrated community-based conservation model, which is now being replicated in other much larger chimpanzee habitat areas elsewhere.

Establishing good relations with people in the community and in the local government is essential for ensuring long-term, effective, conservation efforts in a given area. Such relationships take time to build, and long-term research provides a good basis for building up trust and an understanding that good conservation practices, which benefit wildlife, can also benefit people in the surrounding areas (Kasenene and Ross, Chapter 10). Only then will it be possible to hand over responsibility to the appropriate local authorities.

Long-term research studies also provide a good basis for involving young people and for training both youth and adults in conservation techniques – and this, of course, is essential to the success of conservation in the long run.

REFERENCES

Bygott, D. (1992). *Gombe Stream National Park*. Arusha: Tanzania National Parks.

Goodall, J. van Lawick- (1965). New discoveries among Africa's chimpanzees. *National Geographic Magazine*, **128**, 802–831.

Goodall, J. van Lawick- (1971). *In the Shadow of Man*. London: Collins.

Goodall, J. (1986). *The Chimpanzees of Gombe: Patterns of Behavior*. Cambridge, MA: Belknap Press.

Goodall, J. (1999). *Reason for Hope: a Spiritual Journey*. New York: Time Warner Books.

Greengrass, E. (2002–2003). Sudden decline of a community of chimpanzees *Pan troglodytes* in Gombe National Park, Tanzania. *African Primates*, **6**, 53–54.

Lonsdorf, E. V., Travis, D., Pusey, A. E., and Goodall, J. (2006). Using retrospective health data from the Gombe chimpanzee study to inform future monitoring efforts. *American Journal of Primatology*, **68**, 897–908.

Lukasik, M. (2002). Establishing a long term veterinary project for free-ranging chimpanzees in Tanzania. *Pan Africa News*, **9**, 13–17.

Massawe, E. T. (1992). Assessment of the status of chimpanzee populations in western Tanzania. *African Study Monographs*, **13**, 35–55.

Meshach, S., Buerki, C., and Waeber, P. (2007). Inquiry on the impact of Roots & Shoots in the Lugufu Congolese refugee camp in Tanzania. Jane Goodall Institute, Switzerland.

Ogawa, H., Sakamaki, T., and Idani, G. (2006). The influence of Congolese refugees on chimpanzees in the Lilanshimba area, Tanzania. *Pan Africa News*, **13**, 21–22.

Pintea, L. (2006). Land-use planning in Tanzania. *Imaging Notes*, Winter 2006, 16–21.

Pintea, L., Bauer, M. E., Bolstad, P. V., and Pusey, A. E. (2003). Matching multiscale remote sensing data to interdisciplinary conservation needs: the case for chimpanzees in western Tanzania. Remote Sensing Symposium/Land Satellite Information IV Conference and the ISPRS Commission 1: Integrating Remote Sensing at the Global, Regional, and Local Scale, Denver, CO.

Pusey, A. E., Pintea, L., Wilson, M. L., Kamenya, S., and Goodall, J. (2007). The contribution of long-term research at Gombe National Park to chimpanzee conservation. *Conservation Biology*, **21**, 623–634.

Science North (2002). *Jane Goodall's Wild Chimpanzees*. IMAX Film. Sudbury, Ontario: Science North.

Sharp, P. M., Shaw, G. M., and Hahn, B. H. (2005). Simian immunodeficiency virus infection of chimpanzees. *Journal of Virology*, **79**, 3891–3902.

TANAPA Planning Unit (2005). *Gombe National Park General Management Plan 2005–2015*. Arusha: Tanzania National Parks.

TOSHISADA NISHIDA AND MICHIO NAKAMURA

15

Long-term research and conservation in the Mahale Mountains, Tanzania

THE START OF RESEARCH BY JAPANESE SCIENTISTS

Japan's study of great apes began in 1958, when Kinji Imanishi and Jun'ichiro Itani of the Japan Monkey Centre Gorilla Expedition left for Africa in search of a site for the study of gorillas. The research target was later changed to the chimpanzees (*Pan troglodytes schweinfurthii*) of Tanzania because of political disturbances in eastern Congo, and their search continued for 3 years. At four sites in western Tanzania, the researchers used three different methods in attempts to habituate chimpanzees: planting food crops such as banana and sugarcane (provisioning), presenting a tame infant chimpanzee, and making contact with wild chimpanzees without any form of artificial intervention (Nishida, 1990).

In 1965, Nishida tried to attract chimpanzees by planting sugarcane in the Kasoje area, along the western foot of the Mahale Mountains. In March 1966, K group chimpanzees began to visit the plantation, and in 1968 the M group started visiting. Because the feeding area happened to be inside the overlapping areas of the two unit groups ("communities"), it was possible to elucidate the social units among chimpanzees, antagonistic relationships between groups, and the female transfer system (Nishida, 1968; Nishida and Kawanaka, 1972). The Mahale Mountains Chimpanzee Research Project has continued ever since.

CONSERVATION

Nishida realized the need to conserve the chimpanzee habitat as early as 1967. He was disturbed by the felling of trees in the preferred forests of K group chimpanzees and feared that the forests would disappear forever.

In fact, the cultivation cycle of the Tongwe tribe was over a period of 30 to 50 years and their agriculture was sustainable, making coexistence of humans and wild animals possible. Nishida, however, was unaware of the people's traditional ways of life and did not realize how often the same forest was cleared in the cultivation cycle. He asked the acting director of the Game Department, Mr. J. S. Capon, if it would be possible to establish a wildlife reserve at Mahale. As 200 residents lived in the Kasoje area, Capon stated that 200 shillings per person (or £2000) would be necessary for compensation. Nishida, a student, could not afford £2000. By the early 1970s, however, Nishida and his fellow researchers had come to believe that the Tongwe's traditional agricultural system was sustainable, and that humans and chimpanzees could coexist if a system were put in place to prevent future large-scale, commercial exploitation of the forests.

From 1973 to 1974, Itani and Nishida consulted with officials of the Ministry of Natural Resources, and in particular, with the Director of the Game Division, Raphael Jingu, and the Director General of Tanzania National Parks (TANAPA), Derek Bryceson, to determine what kind of wildlife reserve to establish at Mahale, and proposed a wildlife reserve in which humans could harvest small animals and plants to the extent that consumption was sustainable. Tanzanian officials, however, based on their own long-term experience, considered a human–chimp coexistence project to be inappropriate because, in their opinion, conflict between wildlife and humans would be inevitable. The officials therefore rejected the idea of a wildlife reserve and instead proposed a National Park. Although no one wanted old friends driven out of their homes, the researchers realized that a National Park was one way to preserve the wild chimpanzees.

In 1974, Itani and Nishida consulted with the Japanese ambassador to Tanzania regarding how the Japanese government could help establish a National Park. The ambassador explained that Japanese Official Development Assistance (ODA) was only for the promotion of projects that met the basic human needs of developing countries, such as the construction of bridges, ports, roads, and water works. However, he added that it might be possible to send experts to sites to research conservation plans. Nishida and Shigeo Uehara drafted "An Introduction to the Mahale Mountains: Picturesque Mountain Massifs with a Forest of Chimpanzees" and proposed the plan to the Ministry of Natural Resources and Tourism. The Japanese government accepted the proposal by the Tanzanian government, and in 1975 the Japan International Cooperation Agency (JICA) began to send primatology experts to Tanzania. This continued until

1988. Researchers collected the basic ecological data on chimpanzees from fauna and flora to ranging behavior, feeding ecology, and demography (Anonymous, 1980). During this period, the Serengeti Wildlife Research Institute (currently the Tanzania Wildlife Research Institute) established the Mahale Wildlife Research Centre in 1979. Erasmus Tarimo, the first head of the Centre, made an effort to prepare for the establishment of the Park by initiating the land survey and cutting the Park boundary.

Itani and Nishida had proposed the establishment of a "foot-walking" National Park instead of the automobile parks that are ubiquitous in Tanzania, because the terrain of Mahale is steep and road construction would have destroyed fragile environments such as the Sinsiba Swamp. The Tanzanian National Parks (TANAPA) accepted the idea. However, the development of a National Park takes time; the researchers waited and waited, and even organized an international letter-writing campaign to accelerate the process. From July 1980 to February 1981, Irwin Bernstein, Robin Dunbar, Ardith Eudey, Stephan Gartlan, Birute Galidikas, Annie Gautier, Barbara Harrison, Robert Hinde, Donald Lindburg, Bill Mason, Bill McGrew, Russ Mittermeier, John Oates, Geza Teleki, Richard Wrangham, and many others signed a petition and wrote to the Prime Minister of the Japanese government. At the same time, Itani and Nishida raised 18 million yen in a fundraising campaign, donating 13 million yen to TANAPA. The remaining money was used to establish a library and guest houses.

It took many years for the village, ward, area, and regional development committees, TANAPA, and the Ministry and Parliament to reach a consensus on the Park. However, in 1985, President Julius Nyerere finally designated the Mahale Mountains as a National Park. The Mahale Wildlife Research Centre was engaged in the Park's business including the collection of entrance fees and patrolling until the first Chief Park Warden, A. H. Seki, arrived in 1989.

INTRODUCTION OF ECOTOURISM

In 1989, Irishman Roland Purcell introduced an ecotourism project similar to the mountain gorilla tour operated at the Karisoke Research Center. Up to six visitors at a time were accepted into the tented camp and allowed to spend up to 1 hour observing the chimpanzees. Wealthy visitors, primarily from the United States, the United Kingdom, and Germany, were transported via light aircraft from Nairobi or Arusha to Mahale. Purcell and his wife Zoë were extremely considerate and careful not to disturb the chimpanzees or the researchers. The visiting crew came and left so inconspicuously that sometimes the researchers did not even notice them. Thus, ecotourism began at Mahale without any conflict with ongoing research. In the mid 1990s, TANAPA began to accept tourists in its own Banda tourist camp; these visitors are generally younger and not as wealthy as Purcell's, and they tend to arrive at the park via the steamship *Liemba*.

COMMERCIAL TOURISM FLOURISHES

In the early 2000s, two more companies joined the tourist business, increasing the number of tourists to the Park from only 200 in the early 1990s to more than 1000 in 2005 (Fig. 15.1). Considering the total size of the Park (1614 km^2), 1000 annual visitors may appear to be a small number, averaging only three visitors per day. However, visitors are concentrated in a tiny part (<20 km^2) of the range (30 km^2) inhabited by the habituated chimpanzees, and most come during the dry season from July to September. Even if a tourist group has only six guests, it is accompanied by a park ranger, chimp trackers, a tourist guide, and a manager. The same visitor groups often visit chimpanzees for 2 or 3 consecutive days, and often two or more groups from different tourist companies visit the same place. Furthermore, researchers and research assistants follow the same chimpanzee groups that the tourists come to observe, thus finding themselves in the same places. The result is that sometimes more than 20 people surround two or three chimpanzees. Sometimes a filmmaker joins the scene, and consequently, some of the watchers, researchers, tourists, and filmmakers become frustrated and begin to resent one another.

PREVALENCE OF INFECTIOUS DISEASE

Female chimpanzees with small infants often run away from the observation path or hide themselves in the bush to avoid approaching tourists. One day, Nakamura was following the M-group-born (thus very tame) mother Abi in the bush. Abi suddenly stopped and turned away before reaching an observation trail ahead of her. Nakamura then noticed a tourist group on the trail. Abi stayed quietly in the dense bush with her baby until the tourist group had left in search of a different chimpanzee party. Then Abi proceeded in the original direction and crossed the trail. Abi had detected the presence of the tourists and had decided not to come into contact with them. In this case, the tourist group did not notice Abi in the bush, but if they had, and had followed her they might have chased her away.

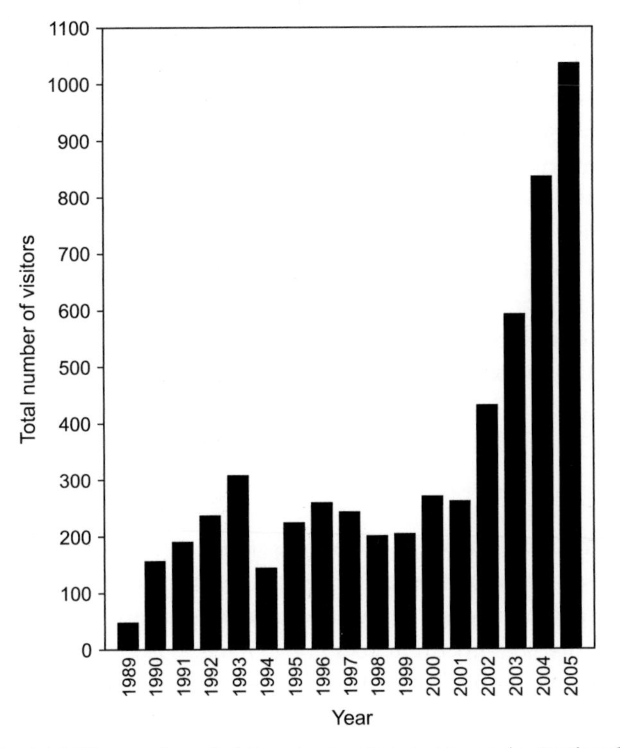

Fig. 15.1. The number of visitors to the Mahale Mountains National Park between 1989 and 2005.

Thus, not only do natural grouping patterns deteriorate (which is undesirable from the viewpoint of scientific research) but also chimpanzees seem to feel constant stress from overcrowding. Although we do not know how overcrowding influences chimpanzees' health, it is definitely possible that stress weakens their immune systems. To investigate this issue, Shiho Fujita began to collect feces during both the high and low tourist seasons to compare the stress hormone content.

Researchers have noticed chimpanzees coughing and sneezing, exhibiting runny noses, loss of appetite, and lethargy (lying on the ground) at certain times in almost every year. The chimpanzees do not always die of this flu-like disease, but the disease did kill 11 chimpanzees in 1993 (Hosaka, 1995) and 12 chimpanzees in 2006 (Hanamura *et al.*, 2008). The "outbreaks" (as operationally defined when more than ten chimpanzees are observed coughing within a month) occur in the dry season between June and October (Nishida *et al.*, 2007). Researchers have assumed that some of the viruses are being transmitted from humans to chimpanzees, which is consistent with the fact that the outbreaks coincide with the high tourist/research season. The numbers of tourists and researchers are greatest in the dry season. Veterinarians from the Japan Monkey Center–Kyoto University team and from the Virginia Polytechnic Institute and State University have begun to acquire non-invasive collections of feces, urine, food wadges, mucus, blood, and other matter from the chimpanzees to identify pathogens causing the flu-like disease. They also collect materials from human subjects (researchers, tourists, research assistants, and villagers) to compare pathogens. For the time being, all observers, researchers, tourists, and park guides must wear face masks while in close proximity (≤10 m) to chimpanzees (Hanamura *et al.*, 2006).

RESTORING ECOTOURISM

Ecotourism is often merely a catchphrase (Mugisha, Chapter 11); commercial tourism is flourishing at Mahale. The best way to restore ecotourism would be to limit licenses to only one travel agency other than TANAPA's Banda. Business competition does not go hand in hand with ecotourism. Competition between tourist agencies results in better services for tourists, including longer observation hours and closer access to chimpanzees, both of which are hazardous to the chimpanzees. Even if it is impossible to grant a license to just one tourist company, TANAPA should organize an integrated booking system so that only one agent can send a single tourist group to the bush at a particular time of day.

While park authorities need money for park management and community conservation, keeping the number of tourists to a minimum and taking in a lot of income appear incompatible, although a solution does exist; a high entrance fee to the park could be introduced as was done in Rwanda. Alternatively, it would not be necessary to raise the general park fee if two separate fees were established: an entrance fee and an observation fee. In this way, a tourist could enter the park by paying, for example, $25 per day, but to observe the chimpanzees for 1 hour would cost an additional $100. If this system worked, visitors could stay for 3 days and pay a total of $175 to observe the chimpanzees once and spend the remainder of their time in pursuit of other entertainment such as swimming and taking walks. Thus, many kinds of activities should be encouraged, from bird and monkey watching to mountaineering, swimming, diving, and fishing. A seasonal fee system might also be introduced to encourage visits during the rainy season (Nishida, 2005).

STATUS OF UNHABITUATED CHIMPANZEES IN AND AROUND MAHALE

Currently, the only habituated group of chimpanzees at Mahale is the M group, which is one reason for the overcrowding of people around this group. To alleviate the burden on the M group, Fumio Fukuda conducted habituation efforts with another chimpanzee group, the B group, from 1994 to 1997, with the financial support of JICA. Due to steeper terrain, wider ranging areas, and suspension of the JICA project, these efforts were unfortunately in vain.

In the late 1990s, it was noticed that another unit group (the Y group) was using the former range of the K group, which had become extinct in the 1980s (Nishida *et al.*, 1985), and between 2001 and 2005 Mitsue Matsuya tried to habituate this group. Since 2005, Tetsuya Sakamaki and Michio Nakamura have been working on the survey and habituation of this group (Sakamaki and Nakamura, 2007; Sakamaki *et al.*, 2007). Habituation is still some time away, but if success is achieved in habituating a second group, the effects of too much human contact on the M group could be ameliorated. If the Y group is used for tourism and the M group for research, research and tourism could be separated as has been done in Uganda (Mugisha, Chapter 11). This would be a welcome development at Mahale.

In addition to detailed behavioral studies on the habituated M group, Mahale researchers have also conducted extensive surveys in and around Mahale National Park (Nishida, 1990; Nakamura and Fukuda, 1999; Shimada, 2003; Zamma *et al.*, 2004; Nakamura *et al.*, 2005). Some chimpanzee groups in the eastern and southern parts of the Park live in much drier areas, where forests are burned by uncontrolled bush fires. Some consider the periodic burning of the Miombo Woodlands to be inevitable; fire promotes regeneration of the undergrowth, which attracts many herbivores. However, researchers have not yet assessed the impact of such fires on chimpanzees, and it appears that such burning might not be very pleasant for forest-dwelling species such as chimpanzees.

It should also be noted that circumstances are much more serious outside the National Park. Large numbers of refugees exist in Tanzania, and some refugee camps have been built in chimpanzee habitats (Nishida *et al.*, 2001). Congolese refugees not only cut down trees for cultivation, housing, and firewood, but they also sometimes hunt for animals, including chimpanzees, as in their culture it is not taboo to eat them. Ogawa and colleagues (2006) warned that destruction by Congolese refugees has become a serious threat to the survival of chimpanzees in the Lilansimba area, where two such refugee camps are located.

OTHER PROBLEMS

In the late 1980s, we noticed the invasion of a foreign tree species, *Senna (Cassia) spectabilis*, in some parts of the M group range (Turner, 1996). *Senna* is a fast-growing tree of South American origin (Watkins, 1960) that flourishes through allelopathy (that is, by killing indigenous plants through toxins). The species was introduced originally to western Tanzania as ornamental or shade trees. It was planted in Mahale villages and the research camp in 1967. *Senna* is not fire resistant, and its saplings are burned to the ground during the dry season every year. However, after the ban on bushfires in 1975, the trees began to extend beyond Kasiha village and the Kansyana research camp. Some experts have warned that *Senna* could take over most of the Park area and destroy plants edible to chimpanzees (Ruho, cited in Nishida, 1996a). Since the early 1990s, systematic girdling and cutting of *Senna* trees (Wakibara, 1998) has been only partially successful. However, Nishida is optimistic because the expansion of *Senna* has not reached the old secondary forest, and saplings of important food plants such as *Saba* and *Pseudospondias* have been thriving in the *Senna* forest (T. Nishida, unpublished observation).

A more serious problem is the existence of a huge fishing village inside the Park boundary (Nishida *et al.*, 2001). Although most Tongwe residents moved out of their territory on the Mahale peninsula with little protest, some immigrants from Zaire (Democratic Republic of the Congo) engaged in *dagaa* (a sardine-like small fish) fishing strongly resisted leaving the Park zone. After 10 years of dispute, they somehow succeeded in pushing the Park boundary away from the fishing village. This solution is problematic because the newcomers remain within the Park at the expense of long-term residents who had a traditional sustainable lifestyle with the natural environment and upon whom chimpanzees and other wildlife depended for their survival, hence the establishment of the National Park (Nishida, 1996b). The fishing village has gradually begun to erode the *Brachystegia* woodland to the east (Fig. 15.2).

MAHALE WILDLIFE CONSERVATION SOCIETY

In 1994, together with Hosea Y. Kayumbo of the University of Dar es Salaam and Toshimichi Nemoto of Japan–Tanzania Tours, Ltd., researchers established a Tanzanian non-governmental organization, the Mahale Wildlife Conservation Society. Kapepwa I. Tambila and John Mitani joined the researchers as founding members at the inaugural meeting held at

Fig. 15.2. A village inside the National Park boundary. Trees are sparse due to cutting and burning.

the University of Dar es Salaam (Nishida, 1995). The mission of the conservation society is to conserve the natural environment of western Tanzania (and of chimpanzees in particular) and to raise public awareness regarding the importance of natural conservation. Activities have included the publication (Volume 1 in 1994) of the newsletter *Pan Africa News*, http://mahale.web.infoseek.co.jp/PAN/index.html, a forum through which researchers of the genus *Pan* (chimpanzees and bonobos) can exchange ideas for research and conservation; the convening of an international conference commemorating 30 years of research at Mahale in 1995; the initiation of the *Senna* eradication project; the building of a primary school at Katumbi village, to which most of the former villagers of Kasoje (which overlaps with the M group range) moved upon the establishment of the Park; the construction of a visitor center at the Park headquarters at Bilenge; and the inviting of primary school students to the Park. The activities of the Mahale Wildlife Conservation Society have been aided financially by JICA, the Japan Environment Corporation and Japanese Embassy, in addition to the receipt of donations from society members.

Kenji and Hatsuko Kawanaka raised money for a fund called the Watoto Foundation and provided supplies and school fees to the children of former residents between 1999 and 2003.

SUMMARY

Long-term research is important from the viewpoints of conservation, science, and culture, and helps the local economy by providing employment to residents. Every day, researchers monitor the activities, health, and demographic changes of a population of habituated wildlife, such as the chimpanzees. Thus, researchers can warn the public and government immediately when a crisis such as poaching or an epidemic occurs. Through new findings, such as the close behavioral similarity of humans and chimpanzees, long-term research creates new value for wildlife and helps to establish wildlife reserves such as National Parks. In this way, research can bring about new business, such as ecotourism, through which residents use natural resources in a sustainable manner, at the expense of direct, short-term exploitation of these resources. However, the phrase "easier said than done" applies to ecotourism. Ecotourism easily leads to commercial tourism and can even destroy the natural resources by attracting too many people to a single park area. Longsightedness, frugality, care, flexibility, and cooperation are necessary to ensure the success of ecotourism, just as in research. If there had been no long-term research at Mahale, there would have been no Mahale Mountains National Park.

ACKNOWLEDGMENTS

We thank the Tanzania Commission for Science & Technology, Tanzania National Parks, the Tanzania Wildlife Research Institute, the Mahale Mountains National Park and Mahale Mountains Wildlife Research Centre for permission for research, advice and logistic support; the Ministry of Education, Culture, Sports, Science & Technology, Japan; the Ministry of the Environment; the Japan International Cooperation Agency; the Japan Environment Corporation and the Ministry of Environment for support of our long-term project. This study was financially supported by the Global Environment Research Fund F-061 to TN and MEXT Grant-in-Aid for Scientific Research 16255007 & 19255008 to TN.

REFERENCES

Anonymous (1980). *Mahale*. Tokyo: Japan International Cooperation Agency.

Hanamura, S., Kiyono, M., Nakamura, M. *et al.* (2006). A new code of observation employed at Mahale: prevention against a flu-like disease. *Pan Africa News*, **13**, 13–16.

Hanamura, S., Kiyono, M., Lukasik-Braum, M., Mlengeya, T., Nakamura, M., and Nishida, T. (2008). Chimpanzee deaths at Mahale caused by a flu-like disease. *Primates*, **49**, 77–80.

Hosaka, K. (1995). Epidemics and wild chimpanzee study groups. Mahale: a single flu epidemic killed at least 11 chimps. *Pan Africa News*, **2**, 3–4.
Nakamura, M. and Fukuda, F. (1999). Chimpanzees to the east of the Mahale Mountains. *Pan Africa News*, **6**, 5–7.
Nakamura, M., Nishie, H., and Mwinuka, C. (2005). Survey of the southern part of the Mahale Mountains. *Pan Africa News*, **12**, 5–8.
Nishida, T. (1968). The social group of wild chimpanzees in the Mahali Mountains. *Primates*, **9**, 167–224.
Nishida, T. (1990). A quarter century of research in the Mahale Mountains: an overview. In *The Chimpanzees of the Mahale Mountains: Sexual and Life History Strategies*, ed. T. Nishida. Tokyo: University of Tokyo Press, pp. 3–35.
Nishida, T. (1995). Inauguration meeting of Mahale Wildlife Conservation Society held at the University of Dar es Salaam. *Pan Africa News*, **2**, 1–7.
Nishida, T. (1996a). Eradication of the invasive, exotic tree *Senna spectabilis* in the Mahale Mountains. *Pan Africa News*, **3**, 6–7.
Nishida, T. (1996b). Thirty years of chimpanzee research at the Mahale Mountains National Park. In *Proceedings of a Scientific Seminar to Mark 30 Years of Chimpanzee Research in the Mahale Mountains National Park*, Dar es Salaam, December 4–6, 1995. Arusha: Serengeti Wildlife Research Institute, pp. 7–17.
Nishida, T. (2005). Introduction of seasonal park fee system to Mahale Mountains National Park: a proposal. *Pan Africa News* **12**, 17–19.
Nishida, T. and Kawanaka, K. (1972). Inter-unit-group relationships among wild chimpanzees of the Mahali Mountains. *Kyoto University African Studies*, **7**, 131–169.
Nishida, T., Hiraiwa-Hasegawa, M., Hasegawa, T., and Takahata, Y. (1985). Group extinction and female transfer in wild chimpanzees in the Mahale National Park, Tanzania. *Zeitschrift für Tierpsychologie*, **67**, 284–301.
Nishida, T., Wrangham, R. W., Jones, J. H., Marshall, A., and Wakibara, J. (2001). Do chimpanzees survive the 21st century? *Conference Proceedings of the Apes: Challenges for the 21st Century*. Chicago: Chicago Zoological Society, pp. 43–51.
Nishida, T., Fujimoto, M., Fujita, S. *et al.* (2007). Prevalence of infectious diseases among the chimpanzees of Mahale. *Primate Report*, **23**, 5–6.
Ogawa, H., Sakamaki, T., and Idani, G. (2006). The influence of Congolese refugees on chimpanzees in the Lilanshimba area, Tanzania. *Pan Africa News*, **13**, 21–22.
Sakamaki, T. and Nakamura, M. (2007). Preliminary survey of unhabituated chimpanzees in the Mahale Mountains National Park, Tanzania: behavioral diversity across neighboring unit-groups and intergroup relationships. In *Formation of a Strategic Base for Biodiversity Studies*. Kyoto: The 21st Century COE Program of Kyoto University, pp. 278–280.
Sakamaki, T., Nakamura, M., and Nishida, T. (2007). Evidence of cultural differences in diet between two neighboring unit groups of chimpanzees in Mahale Mountains National Park, Tanzania. *Pan Africa News*, **14**, 3–5.
Shimada, M. (2003). A note on the southern neighboring groups of M group. *Pan Africa News*, **10**, 11–14.
Turner, L. A. (1996). Invasive plant in chimpanzee habitat at Mahale. *Pan Africa News*, **3**, 5.
Wakibara, J. V. (1998). Observations on the pilot control of *Senna spectabilis*, an invasive exotic tree in the Mahale Mountains National Park, Western Tanzania. *Pan Africa News*, **5**, 4–6.
Watkins, G. (1960). *Trees and Shrubs for Planting in Tanganyika*. Dar es Salaam: The Government Printer.
Zamma, K., Inoue, E., Mwami, M., Haluna, B., Athumani, S., and Huseni, S. (2004). On the chimpanzees of Kakungu, Karobwa and Ntakata. *Pan Africa News*, **11**, 8–10.

CHRISTOPHE BOESCH, HEDWIGE BOESCH, ZORO BERTIN GONÉ BI,
EMMANUELLE NORMAND, AND ILKA HERBINGER

16

The contribution of long-term research by the Taï Chimpanzee Project to conservation

One day, two poachers were in the forest and entered the research area of the Taï Chimpanzee Project. They knew that many more monkeys and duikers could be found here than in other parts of the park. After a long walk they heard chimpanzee calls. The chimpanzee group moved toward them without any reaction to their presence. The younger poacher, who was there to carry meat, told the older one with the gun to shoot. But the older one came from a village that had been visited by the Wild Chimpanzee Foundation awareness team. "Wait! People in the village say that the chimpanzees are like humans," he answered. "Let's first have a look." It was the Coula nut season and after seeing how the chimpanzees were using hammers to break the nuts open and how some mothers were sharing the nuts they opened with their infants, the older said "They are right in the village. Chimpanzees are like humans. Let's move on." The poachers continued on their way without shooting at the chimpanzees.

This anecdote was told to us during one of the discussions we had in the village. It illustrates nicely how bringing information about the true abilities of chimpanzees to local populations can contribute directly to saving the lives of this highly endangered species. Scientists can play an important role in conservation and should get involved in sharing their knowledge with local people.

INTRODUCTION

Protection of wild animal populations is an increasing worry for the future of our planet. Successfully protecting wildlife takes a complex mix of activities ranging from law enforcement to sustainable development. Getting the blend of activities just right requires detailed knowledge of the conditions at a given site. This kind of knowledge is often difficult for large conservation organizations, which operate internationally, to obtain and prioritize. Consequently, the best and most efficient way of protecting key animal populations has proven to be the constant and long-term presence of research projects. Long-term field research projects following chimpanzees *(Pan troglodytes)* over generations contribute to the survival of their study populations, not only because the presence of researchers directly protects the animals, but also because such projects provide jobs to the local populations and are therefore viewed with sympathy and respect. However, given that populations sometimes decline while being observed, it is also possible that there are some negative demographic effects of long-term research (Köndgen *et al.*, 2008).

In this contribution, we detail some of the complex interactions between a long-term research project in the Taï National Park, Côte d'Ivoire and the local human populations. We also illustrate how long-term projects, by becoming more involved with the social network of the region, can contribute, not only to the protection of chimpanzees, but also to the forest in which they live. Specifically, we will illustrate how educational activities, including a traveling theater group, were able to modify the attitude of people toward chimpanzees in a large region of West Africa and amongst people with differing cultural and religious backgrounds.

HISTORICAL BACKGROUND

The Taï Chimpanzee Project was initiated in 1976 by Christophe and Hedwige Boesch, who were interested in studying nut-cracking behavior in the Taï forest. Two reports had mentioned nut-cracking sites in this forest, and the local population designated the chimpanzees as the nut-crackers. But nobody in the research community appeared to have observed chimpanzees in the act of cracking nuts (Rahm, 1971; Struhsaker and Hunkeler, 1971). After we first witnessed a chimpanzee holding a hammer at a nut-cracking site, the potential for a long-term research project on the Taï chimpanzees became evident (Boesch, 1978).

We started in 1979 with the support of the Swiss Research Foundation, which financed the long-term project without interruption for 18 years. Nut-cracking turned out to be a stroke of luck for a research topic interested in tool behaviour, as pounding nuts is noisy and can be heard over a few hundred meters. In addition, the nuts open with a very distinctive cracking sound so that nut-cracking efficiency data could be collected even without visual contact. Thanks to this lucky situation, initial data were collected before habituation was achieved, guaranteeing the new project continuing financial support.

As the project developed, it rapidly became evident that, like most forests of the continent, the Taï National Park was subject to many pressures. Illegal logging and agriculture were threatening the periphery of the National Park. Logging was conducted within the limits of the National Park, which at the time were not clearly demarcated. In addition, hunting for meat is, in forested regions of Africa, a major source of protein. We were rapidly confronted with the problem of encounters with poachers.

The main research group of chimpanzees was habituated within 5 years and all group members identified 2 years before. After 7 years, the number of individuals started to decline (Boesch and Boesch-Achermann, 2000). Therefore, with the help of students, in 1988 we started to habituate a neighboring group whose territory lay south of the original study group. This process was complex because it was impossible to know the territorial limits of the new group of chimpanzees. After 3 years it turned out that the new group we were trying to habituate was actually two distinct groups. With the help of six students and local assistants, we were able to habituate both groups and, for the first time, three neighboring chimpanzee communities were followed at the same time, and in the same way.

In 1989, we included local young villagers in the project as field assistants. To our delight, it turned out that, once their natural fears of the forest were overcome, they were extremely keen and accurate observers of the chimpanzees. We recruited and trained a team of about 12 local field assistants who followed three communities of chimpanzees on a daily basis. Many of those locals were very careful and skilled observers, so we trained some of them to record daily social interactions of target individuals. Others collected information on infant development.

Now, after 28 years, the Taï chimpanzee project has habituated a fourth group, in response to the fate of one of the habituated groups, the middle group, which had been reduced to one adult male and two adult females. Three chimpanzee groups are still followed on a daily basis, and the data collected by the field assistants have been entered into a large database that allows us to follow long-term trends in the social behavior of the chimpanzees. The project now includes genetic, hormonal, and veterinary approaches in addition to the original ecological and behavioral ones, and there has been an increase in the number of

Ivorian students and field assistants working on chimpanzee-related research.

RESEARCH CONTRIBUTION

One major reason for starting the Taï chimpanzee project was to study chimpanzees living in the heart of the rainforest. Previous field projects on chimpanzees, like the Gombe and Mahale chimpanzee projects, were studying individuals living in much more open, mixed habitats, including savanna, woodlands, and gallery forests. This particular habitat was chosen by Louis Leakey, the famous anthropologist who initiated the Gombe chimpanzee project. He wanted observations on the behavior of the chimpanzees to test the theory that our ancestors were adapted to more open habitat and acquired human-like behavior, such as tool use and hunting behavior, after they left the forest (Collins and Goodall, Chapter 14). To complement Louis Leakey's approach to understanding the effect of ecology on chimpanzee behavior, it seemed important to know how chimpanzees behave in the forest. Nowadays, the Taï chimpanzee project remains one of the primary rainforest projects following rainforest chimpanzees. Thus, the initial incentive for the Taï chimpanzee project was comparative and this is still true today.

Taï chimpanzees revealed themselves as astonishing representatives of forest chimpanzees. Contrary to all predictions of the "out-of-savanna model," which assumes the move to open habitat favored the apparition of "human-like traits" such as tool use, hunting, and cooperation, it turned out that Taï chimpanzees used more tools, made them in more varied ways, hunted more in groups, and used more elaborate cooperative techniques than the chimpanzees living in more open habitats (Boesch-Achermann and Boesch, 1994; Boesch and Boesch-Achermann, 2000). This has forced the anthropological community to reconsider the "out-of-savanna model" and especially to look more critically at the effect the environment might have had on our early ancestors.

HUMAN POACHING PRESSURE

Most forested regions of Africa are quite isolated, often accessible only by unpaved roads. Consequently, one reliable way for humans living in those regions to get access to meat protein is by hunting wild animals. This is more common in regions where people still have a traditional animist religion, and less so in Muslim regions, like Uganda or Tanzania in East

Africa. Hunters make small temporary camps in the forest and hunt mostly monkeys during the day and forest duikers at night. They smoke the meat over a fire in the camp until they have accumulated enough to take back to their village to sell. This traditional hunting nowadays is done with guns, which can target prey precisely high in the canopy and are very quickly recharged. Thus, it is not difficult for one hunter to kill six to eight monkeys from one group within minutes. Red colobus (*Procolobus rufomitratus*) monkeys are the most abundant monkeys in many African forests, but they are rather noisy and slow to run away from threat (as males tend to face danger as a group). Such a reaction is very effective at repelling eagles and chimpanzees, their natural predators, but totally maladapted when facing a gun. The consequence is that red colobus monkeys have been totally hunted out in many African forests. The "empty forest syndrome" in which forest trees stand intact but the fauna has been extinguished is a dramatic illustration of the radical effect of hunting.

We quickly detected many such hunting camps in the Taï forest and discussed the issue with the villagers living near our research camp. All agreed that hunting was not allowed in the Park and that these people should not be there, but no one knew who they were. We adopted an attitude of informing the Park guards systematically of the presence of poachers and of destroying the remains of such temporary camps whenever we encountered them within the research area. Sadly, we sometimes also found snares placed in the forest to trap forest duikers. Although we systematically removed all snares we encountered, occasionally the chimpanzees would get caught in them accidentally and we saw one dying from the infection it produced. Park guards came to support us and to look for poachers on many occasions and news spread that researchers were working in the forest. Poachers became more careful and avoided the research area.

Despite the decrease of poaching, it was never eradicated and remains a threat to the chimpanzees. Due to the distance of our research area from the villages, snares remain rare and most of the poaching is done with guns. The main victims are monkeys, especially the red colobus monkeys. But, sadly, chimpanzee meat stands very high on the preference list (see below), and poachers try their luck whenever they encounter chimpanzees.

Recently, another problem has appeared. Because of the effective repellent effect of our constant presence in the forest, monkey densities were higher in our study area than in other parts of the forest. Poachers have started to realize that they can kill more monkeys with less effort inside the research area than elsewhere. This has become a strong incentive for poachers to enter our research area and has forced us to bcome engaged more actively with education, explaining to the local populations the benefit of protecting one of the last intact regions of the Park (see below).

If poaching has devastating effects on monkey and duiker populations, the effect on chimpanzees is less well known. Surveys in Gabon have suggested that a decrease in chimpanzee population of up to 95% can result from living in non-Protected Areas where poaching pressure and other human encroachments are high (Tutin and Fernandez, 1984). In the Taï project, we obtained clear evidence that eight chimpanzees we studied for years fell victim to poachers. We know that two of them were eaten later in local villages and that one baby was sold in the capital, Abidjan. Six of them were killed in November 2002, during the most chaotic period of civil unrest in the country. This illustrates how temporary disorganization in the civil legal structure of a country can have a dramatic impact on conservation. We suspect that three more chimpanzee disappearances were due to poachers. We also know of two individuals that were shot but survived their injuries, as we found shotgun pellets lodged in their bones when they died years later from natural causes.

This recurrent problem of poaching of chimpanzees contributed to our decision, in 2000, to create the Wild Chimpanzee Foundation (www.wildchimps.org), an international non-governmental organization aimed specifically at protecting wild chimpanzee populations in areas where they have some prospect of surviving. The foundation became immediately active since we saw how rapid and how devastating human encroachments into the forest could be.

CONSERVATION THROUGH THE WILD CHIMPANZEE FOUNDATION

The Wild Chimpanzee Foundation (WCF) resulted from our direct experience in the forest and therefore adopted three main tenets as a founding philosophy.

- Scientifically based evidence must guide WCF activities if we want to be quick and efficient in helping both wild populations and the forested habitats necessary to their survival.
- Capacity building is imperative for gaining the support of local populations on a long-term basis and for ascertaining the sustainability of long-term survival of wild populations and ecosystems.

- "Now or never" guides our actions: money, actions and people working on the ground with concrete projects must contribute as directly as possible to the survival of the chimpanzees throughout Africa.

Following this philosophy and considering the fact that we are a new and relatively small organization, we decided that filling gaps left by the activities of larger conservation NGOs would be directly relevant to the survival of the apes. As a result of our analysis, we initiated activities in three main domains.

- Environmental education in the villages inform local people about their environmental heritage. To capture the most interest and rely on oral tradition, so important throughout Africa, we established a traveling theater company that visits villages located near key chimpanzee populations.
- Scientifically based fauna and human encroachment surveys are very important tools in increasing the effectiveness of conservation activities, and are critical to evaluating the effectiveness of such programs. By providing precise maps on chimpanzee and other animal distributions as well as locating human activities, park managers and surveillance teams are able to react very precisely and accurately to any situation in a Park.
- Supporting local development actions brings employment into the villages which can directly counter illegal activities that are mainly engaged in by young people. Scientists, game wardens, local people, and authorities are helping to target such alternatives more effectively.

Below, we detail the WCF education activities in the villages and show some of the results to illustrate how knowledge built up during a long-term study can improve the message we bring to villagers. We also show how this can lead to changes in the attitude of the local populations towards the chimpanzees.

The theater play

The message of the play was: "The chimpanzees are our cousins in the forest, do not kill them." People are attracted instinctively by theatric representations of real life and especially in societies with oral traditions. WCF decided to use this type of communication in the villages located near Protected Areas with key chimpanzee populations. Theater seemed a promising way to improve the perception of the chimpanzee and to address the issue of its coexistence with the local human population. We worked closely with local companies in Abidjan, Freetown, and Conakry. The piece described the coexistence of chimpanzees and humans and was developed to be convincing and attractive to the local populations.

The structure of the play was a mixture of monolog, dialog, mimes, dances, music, and songs. Dancers, musicians, and other people living in the target region were repeatedly consulted in order to select music and songs that local people could appreciate and even recognize. Each scene of the play, which lasted about 45 minutes, included lyrics and dances that reiterated the message. The official language of the country was used, since each village is often a composite of people from different origins and different language groups. However, phrases in local dialects were embedded in the play, since actors in the company came from the region. Once the play was over, representatives of WCF (Ilka Herbinger, Zoro Bertin Goné Bi, scientists who studied Taï chimpanzees, and Grégoire Nohon, a long-term field assistant in the Taï chimpanzee project) would lead a discussion with the public, answering questions raised by the drama.

We presented the play in over 40 villages around the Taï and Marahoué National Park, in 40 villages around the Gola Reserve Forest in Sierra Leone, in 19 villages in the Fouta Djalon and in 15 villages in the forested region of Guinea targeting an estimated 100 000 spectators in the last 3 years. A video clip of the theater performance can be seen on the WCF webpage: www.wildchimps.org/wcf/video/WCF01_wmv.wmv.

To measure the impact of the theater, we conducted independent sociological studies before the education campaigns visited the villages and 3 months after. These studies were performed by teams of sociologists from the universities of Abidjan, Freetown, and Conakry (Boesch *et al.*, 2007). They did their evaluation totally independently of the WCF awareness team and neither participated in the creation of the play nor in the performance in the villages. During these sociological studies, personal interviews, based on a predefined list of questions, were carried out on a representative sample of the total population in a minimum of 50% of the villages where the play was performed. Here, we will report on the results of these studies, from a total of 480 interviews, around the Taï and Marahoué National Park in Côte d'Ivoire and in the Fouta Djalon in Guinea (Herbinger and Boesch, 2006). The villages were selected to represent the socioeconomic characteristics of the region around Taï and Marahoué National Park and in the Fouta Djalon.

Impact of the theater performances

In general, 95% of the people thought the theater play was a good way of presenting a problem because it reached a large audience. The play was easy to understand, faithful to the reality, and allowed an empathy with the actors. The reception of the play was very positive and 80% of the spectators agreed that the productions represented real situations.

Traditional knowledge and perception of the chimpanzees

In all the different productions, spectators learned and retained detailed information about the behavior of chimpanzees (Fig. 16.1 (a)–(d)): aspects of their social life, their reproductive patterns, and even more their tool use and drumming behavior were remembered accurately 3 months after the play.

Besides the fact that chimpanzees live in groups, traditional knowledge of the behavior of chimpanzees was very limited. For example, before the campaign, fewer than 15% of the people in all three sites could explain how chimpanzees produce drumming sounds. A large majority of people thought it was something mystical. Three months after the theater performance, knowledge had clearly increased in all villages, independent of the differences that existed between them.

Finally, in Guinea, we asked some more questions that we had not considered in the earlier studies in Côte d'Ivoire. Before the theater performance, fewer than 10% of the interviewed people knew that protection of chimpanzees can have positive impacts on the environment. After the campaign, nearly 30% knew that chimpanzees disperse seeds and hence contribute to reforestation. Furthermore, before the campaign, only 17% of the people interviewed considered domestication as harmful or illegal, whereas, after the campaign, nearly 70% knew about the negative impacts of domestication. An even lower percentage (about 10%) knew about the laws protecting chimpanzees, whereas after the campaign, nearly 80% not only knew about the existence of these laws but also could explain their content. An important consequence of their limited knowledge of chimpanzees was that people had a rather negative perception of them. This negative attitude towards chimpanzees dominated in the two regions of Côte d'Ivoire where they were judged to be stupid by over 80% of the respondents in the three regions (see Fig. 16.2 (a), (b), (c)). Thus, the theater performance had an important positive influence, as following the play, chimpanzees were judged to be intelligent by the majority of the respondents in all villages in Côte d'Ivoire as well as in Guinea (Fig. 16.2 (b), (d)).

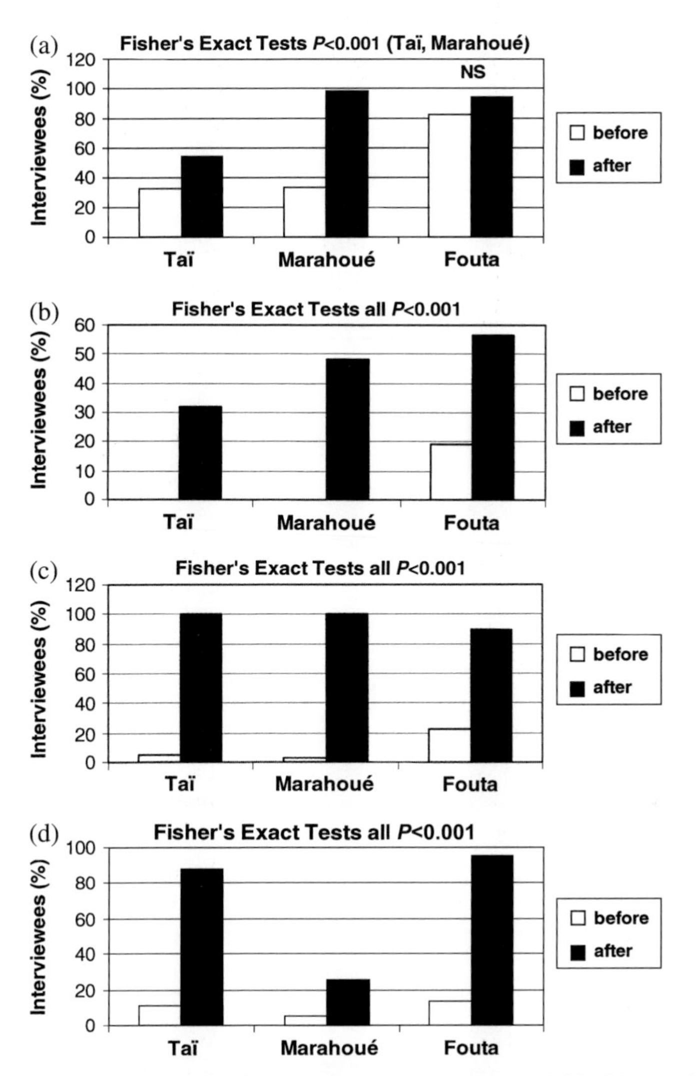

Fig. 16.1. Acquisition of knowledge concerning (a) social life, (b) reproductive patterns, (c) tool use, and (d) drumming behavior through the theater play in two regions of Côte d'Ivoire (Taï and Marahoué NP) and one in Guinea: proportion of interviewees giving positive answers before and 3 months after the theater performance in their villages.

CHANGES BY THE THEATER PLAY IN THE REACTION TOWARDS THE CHIMPANZEES

Did the changes introduced by the plays also affect the way the people reacted to the chimpanzees? Traditionally, people either kill

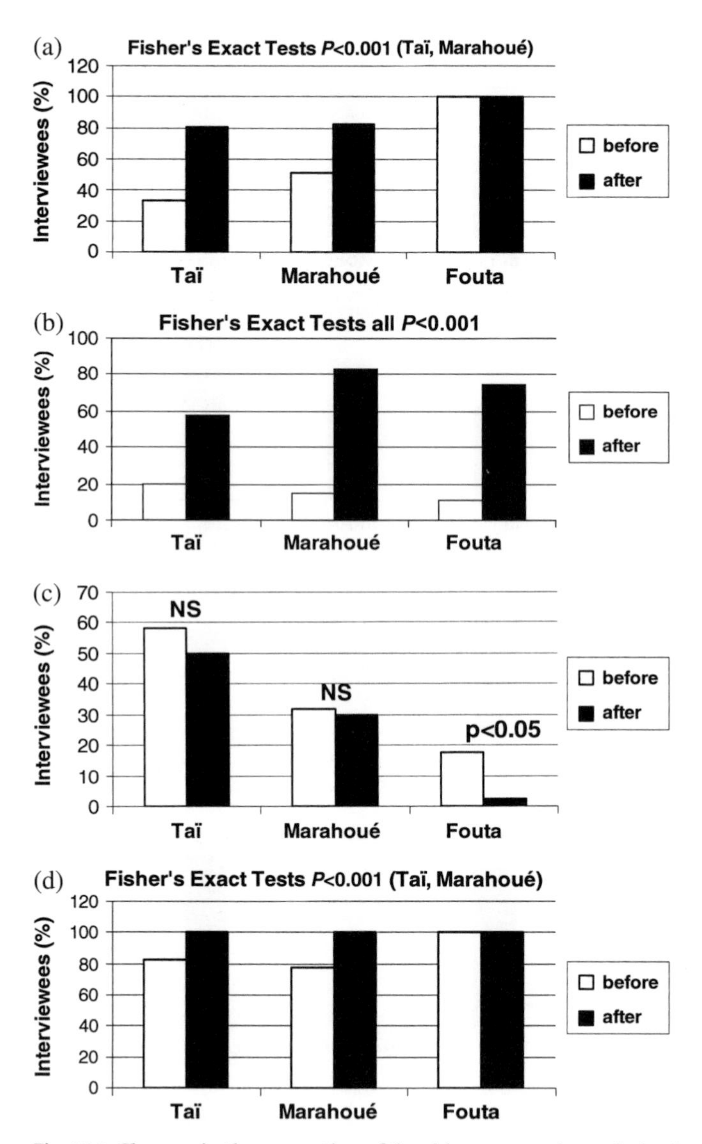

Fig. 16.2. Changes in the perception of the chimpanzees through the theater play in two regions of Côte d'Ivoire (Taï and Marahoué NP) and one in Guinea: (a) positive attributes for chimpanzees, (b) recognized intelligence, or (c) aggressiveness, and (d) readiness to protect chimpanzees.

chimpanzees that enter their fields or chase them away (Fig. 16.3 (a), (b)). People in Côte d'Ivoire were much more inclined to kill them than in Fouta Djalon where people mostly chased them away from their field; 3 months after the theater performance people in both sites were more likely to chase

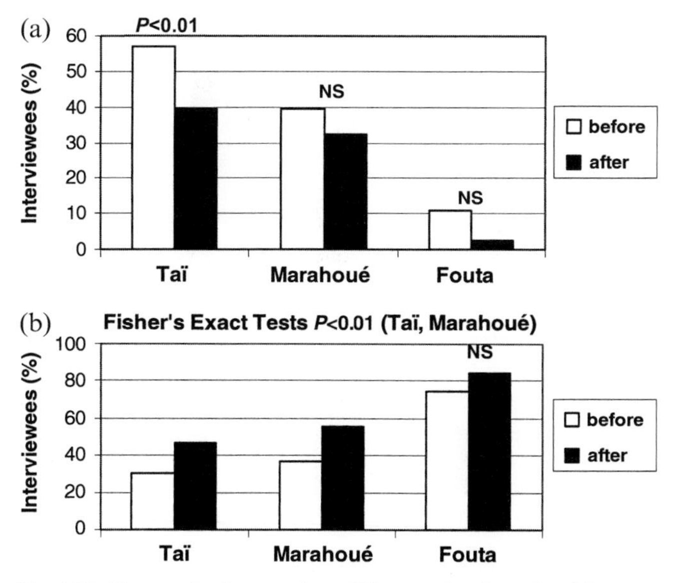

Fig. 16.3. Changes in the reaction of the people when the chimpanzees enter their field through the theater play in two regions of Côte d'Ivoire (Taï and Marahoué NP) and one in Guinea: (a) kill chimpanzees (b) chase them away.

chimpanzees than try to kill them. This change of behavior was greater in Taï, where killing was more frequent before the play (Fig. 16.3 (b)).

How about the consumption of chimpanzees? Because of the prevalence of Islam, people do not eat chimpanzee meat in the Fouta Djalon, Guinea. In the animist regions of Côte d'Ivoire, however, chimpanzees were eaten regularly. The theater performances had a very important effect on the attitude of the local people to the consumption of chimpanzee meat. Traditionally, chimpanzee meat stands very high on the preference list of the people and a large proportion of the villagers declared that they ate chimpanzee meat. Only a minority of the people had totemic relations with the chimpanzees and, therefore, did not eat them (see Fig. 16.4 (a)–(c)).

People that considered them taboo, and did not eat them, often had a traditional connection with chimpanzees (Fig. 16.4 (a)). Usually, some ancestors in their family had had an important positive experience with the chimpanzees, such as having been warned of approaching enemies by the animal or having some family members saved by them, and therefore the family respected them. After the campaign, the percentage of people who consider chimpanzees sacred rose significantly and, spectacularly, some traditional authorities of villages decided to declare the chimpanzee as a taboo for the whole village.

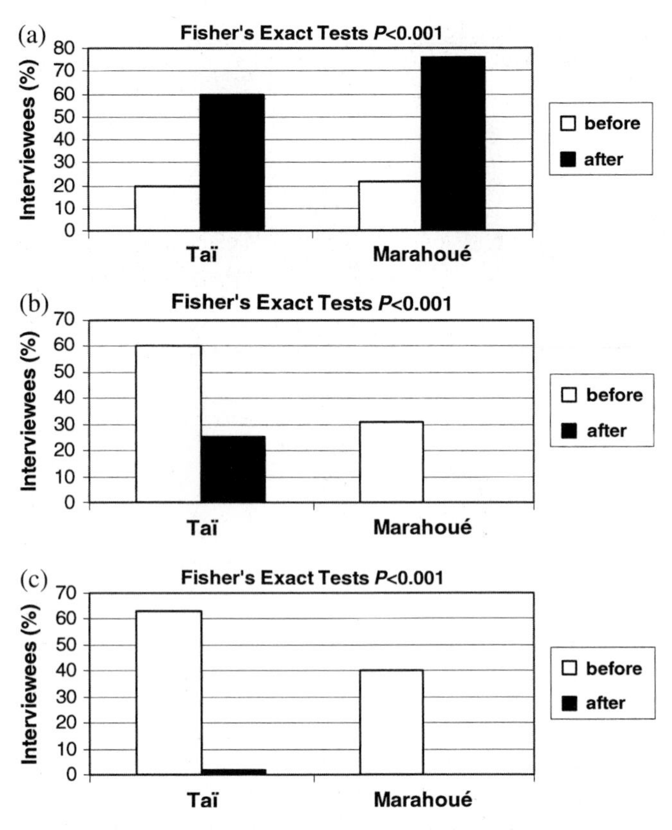

Fig. 16.4. Changes in (a) totemic beliefs protecting Chimpanzees, (b) meat consumption, and (c) preference for chimpanzee meat after the theater play in two regions of Côte d'Ivoire (Taï and Marahoué NP).

Three months after the theater tour, chimpanzee meat moved out of the list of preferred meats like beef, chicken, monkeys, and duikers (Fig. 16.4 (c)). Often, children in the villages called the killing of a chimpanzee murder and the eating of chimpanzee meat cannibalism. Furthermore, we observed an important decrease in the number of people declaring that they still ate chimpanzee (Fig. 16.4 (b)).

SUMMARY

28 years into the Taï Chimpanzee Project, the situation is positive overall, but complex: on one side, many TV documentaries, e.g., BBC "Too Close for Comfort" and "Trials of Life" or National Geographic "The New

Fig. 16.5. Theatrical performance showing a hunter standing over a dead chimpanzee.

Chimpanzee," have brought the impressive scientific information on chimpanzees to the attention of a large audience worldwide. Our constant and active presence on the ground has protected a large area of the Taï National Park in which the chimpanzees and the extraordinary primate community of the park are relatively intact.

On the other side, the human pressure, the rate of deforestation, and the bushmeat trade are a constant throughout the region and more and more chimpanzees and other primates are falling victim to them (Fig. 16.5). Large sections of the Taï National Park have suffered from these problems. This has led to higher pressure on the research area from poachers who reap greater rewards inside this more protected zone.

In other words, the long-term presence of the Taï Chimpanzee Project has preserved the rich biodiversity of the Park and can act as a buffer to repopulate the Park if protective measures are more complete. But such a situation is susceptible to unpredicted changes. In 2002, Côte d'Ivoire went through a period of civil unrest with some fighting and a temporary division of the country into two zones. The project was closed for 6 months due to the security situation, and poaching increased immediately. Six habituated chimpanzees were killed as well as half of the habituated red monkey group of the Taï monkey project (McGraw *et al.*, 2007). Long-term research projects have a very positive influence on conservation but need to be complemented by other measures for such effects to be sustained over longer periods of time.

The long-term research revealed the reality of the gradual chimpanzee decline in the beautiful Taï National Park. Our experience taught us more about poaching organization in the villages, and about the needs of the local population. As a result of some punctual interventions in the villages, we learned what worked, and the WCF grew out of these experiences. Based on knowledge from extensive field experience and date a from important scientific research, the WCF developed a number of projects in a very short time in Côte d'Ivoire. Funding for these activities was quickly found, due to the international recognition gained from our long-term scientific research.

ACKNOWLEDGMENTS

We thank all the following people for having contributed to the long-term data collection of the Taï chimpanzee project that was so important for its continuation: Anderson Dean, Bally Kevin Charles, Bally Louis Bernard, Ban Simone, Bertolani Paco, Betsch Carol, Bolé Camille, Cipoletta Chloe, Crockford Catherine, Dehgenan Olivier, Deschner Tobias, Dji Troh Camille, Donati Franca, Eckhardt Nadin, Gnaha Djirian Appolinaire, Gnombouhou Kouya Gabriel, Gomes Cristina, Goulei Florent, Gouyan Bah Nestor, Guiro Thia Ferdinand, Guy Sylvain, Jensen SivAina, Kohou Nohon Grégoire, Kouakou Yao Célestin, Leendertz Fabian, Lia Denis, Mihi Jean Baptiste, Möbius Yasmin, Néné Kpazahi Honora, N'Guessan Kouamé Antoine, Oulaï Daurid Nicaise, Pahi Tah Alain, Radl Gerhard, Riedel Julia, Schenk Svenja, Sioblo Arsène, Stumpf Rebecca, Tagnon Alphonse, Tahou Mompeho Jonas, Téré Blaise, Wittig Roman, Yro Francois.

We thank all the staff members of the WCF and all partner organizations for performing the awareness program in logistically difficult regions, including the team of Ymako Teatri, the Company Taïbou, all other trained village and school groups, Guinée Ecologie and the sociological teams composed of Acka-Douabele Cinthia, Diallo Baïlo Télivel, Goh Denis. We are grateful to the Guinean authorities and all Ivorian authorities who allowed us to work in the Taï and other National Parks and supported our work in many ways: Ministère de l'Enseignement Superieur et de la Recherche Scientifique, Ministère de l'Environnement et des Eaux et Forêts, Office Ivoirien des Parcs et Reserves en Côte d'Ivoire. We also thank the Centre Suisse de Recherches Scientifiques that has supported the project ever since it was established in Côte d'Ivoire and we thank the Taï Monkey Project for its year-long cooperation. Particular thanks go to the local population in favor of protecting chimpanzees and their habitat. We are most grateful to Burkhardt Alex, Haas Myriam, Lauginie Francis, Levant Christine, Nebel Claudia, and Scheller Marzela for their untiring support in the many tasks of the WCF management and media presentation.

For year-long faithful financial support of the research project, we are very grateful to the Swiss National Science Foundation, the Max-Planck Society and the Robert-Koch Institute. For supporting the activities of the Wild Chimpanzee Foundation, we are most grateful to the Cleveland Metroparks Zoo, Columbus Zoo, Conservation International, Critical Ecosystem Partnership Fund, European Union, Great Ape Conservation Fund of US Fish and Wildlife Service, Great Ape Emergency Conservation Fund, Great Ape Survival Project, Marianne und Benno Lüthi Stiftung, Paul Schiller Stiftung, Great Ape Trust of Iowa, United Nations Environment Program, World Wide Fund for Nature, Zoo Leipzig, Zoo Zürich, Zürcher Tierschutz, the many generous private donors, and the BBC Natural History Unit for allowing us to use the film on Taï chimpanzees in the awareness activities.

We thank Peter Walsh for correcting the English language of this text.

REFERENCES

Boesch, C. (1978). Nouvelles observations sur les chimpanzés de la forêt de Taï (Côte d'Ivoire). *Terre et Vie*, **32**, 195–201.

Boesch, C. and Boesch-Achermann, H. (2000). *The Chimpanzees of the Taï Forest: Behavioural Ecology and Evolution*. Oxford: Oxford University Press.

Boesch, C., Gnakouri, C., Marques, L. *et al.* (2007). Chimpanzee conservation and theater: a case study on an awareness project around the Taï National Park, Côte d'Ivoire. In *Conservation in the 21st Century: Gorillas as a Case Study*, ed. T. Stoinski, P. Mehlman, and D. Steklis. New York: Kluwer Academic/Plenum Publishers.

Boesch-Achermann, H. and Boesch, C. (1994). Hominization in the rainforest: the chimpanzee's piece of the puzzle. *Evolutionary Anthropology*, **3**(Suppl. 1), S9–S16.

Herbinger, I. and Boesch, C. (2006). Interactive theater plays as an effective sensitization tool for the conservation of chimpanzees in West Africa. *Abstracts, 21st International Primatological Society Conference*, Entebbe, Uganda. *International Journal of Primatology*, **27**(Suppl.1).

Hill, K., Boesch, C., Goodall, J., Pusey, A., Williams, J., and Wrangham, R. (2001). Mortality rates among wild chimpanzees. *Journal of Human Evolution*, **40**, 437–450.

Köndgen, S., Kühl, H., N'Goran, P. *et al.* (2008). Pandemic human viruses cause decline of endangered great apes. *Current Biology*, **18**, 260–264.

McGraw, S., Zuberbühler, K., and Noë, R. (2007). *Monkeys of the Taï Forest: An African Primate Community*. Cambridge: Cambridge University Press.

Rahm, U. (1971). L'emploi d'outils par les chimpanzés de l'ouest de la Côte d'Ivoire. *La Terre et la Vie*, **25**, 506–509.
Struhsaker, T.T. and Hunkeler, P. (1971). Evidence of tool-using by chimpanzees in the Ivory Coast. *Folia Primatologica*, **15**, 212–219.
Tutin, C. and Fernandez, M. (1984). Nationwide census of gorilla (*Gorilla g. gorilla*) and chimpanzee (*Pan t. troglodytes*) populations in Gabon. *American Journal of Primatology*, **6**, 313–336.

TETSURO MATSUZAWA AND MAKAN KOUROUMA

17

The Green Corridor Project: long-term research and conservation in Bossou, Guinea

The present chapter describes an ongoing reforestation program known as the "Green Corridor Project" that has grown out of 30 years of research into the Bossou chimpanzee community in Guinea. The Green Corridor Project aims to connect two forested habitats of chimpanzees by planting trees in the intervening savanna. The project was started in 1997 and has provided valuable data on reforesting savanna. It has had some success in extending the range of chimpanzees.

THE CHIMPANZEES OF BOSSOU

Bossou is a village located in the border of Guinea and Liberia, about 1000 km from the capital Conakry. A group of 13 chimpanzees occupies the small forests surrounding the village, into which about 2500 villagers and Liberian and Ivorian refugees are crowded. The core area of the chimpanzees (where they spend most of their time) is about 6 km^2, while their total ranging area is about 30 km^2. The core area consists of very small portions of primary forest, secondary forest, and riverine forest, surrounded by cultivated fields and savanna.

The Bossou chimpanzees were first described by French and Dutch scientists (Lamotte, 1942; Kortlandt, 1986). Since 1976, Japanese scientists and then an international team (KUPRI International) have been studying the chimpanzees at Bossou (Sugiyama and Koman, 1979; Matsuzawa, 2006a). The chimpanzees at Bossou are well known for their use of stone tools. Using mobile stones as hammers and anvils, they crack open the hard shells of oil-palm nuts to eat the kernels (Matsuzawa, 1994, Fig. 17.1). The repertoire of tool use exhibited by chimpanzees of the

Fig. 17.1. Bossou chimpanzees use stone tools and mobile stones as hammer and anvil, to crack open the hard shell of oil-palm nut to eat the kernel (Photo by E. Nogami).

Bossou community is unique (Ohashi, 2006; McGrew, 2004; Yamakoshi, 2001).

An exceptional feature of these chimpanzees is their coexistence with humans and their use of human habitat. Chimpanzees often cross roads using a special sociospatial organization in which adult males guard the females and offspring (Hockings *et al.*, 2006). Males also take papaya fruits from the village and give them to females (Hockings *et al.*, 2007; Ohashi, 2007). People living in Bossou tolerate the chimpanzees because the indigenous ethnic group (the Manon) believe that chimpanzees are their ancestors.

Chimpanzee society is characterized by male philopatry; males stay in the natal community, while females emigrate to adjacent communities. Females at Bossou have their first baby when they are between 9 and 12 years old. During 31 years of recording births at Bossou, there have been four cases where the mother was only 9 years old. This is a little earlier than in other communities (Emery Thompson *et al.*, 2007). All females born into the Bossou community during the study period emigrated out of the community either before giving birth for the first time or immediately afterwards (Matsuzawa, 2006b).

Since 1976 there have been no records of females immigrating into the Bossou community. This might be due to the fact that the community is isolated both by the surrounding human presence and the savanna. Chimpanzees born into the community are able to leave, but non-habituated chimpanzees from other communities cannot easily join the Bossou community. A few males have remained in the Bossou community, so the society still keeps the patrilineal nature common to the genus *Pan*. Among the young chimpanzees, all the females and some of the males have either left or disappeared. The age range of the 12 individuals that left or disappeared was 7–16 years old, with an average age of 10.2 years.

When the author, TM, started fieldwork at Bossou in 1986, the population of the village was about 1500. In 1990 a civil war started in Liberia, and the population almost doubled. After the war the population gradually decreased to the current level of 2500 (A. Kabasawa, personal communication).

The Bossou chimpanzees, like all others across Africa, are being threatened. According to an IUCN report, there are now about 187 000 chimpanzees living in tropical rain forest and surrounding savanna. The species is endangered and the numbers are decreasing.

There seem to be three major threats. The first is deforestation because of pressures from a growing human population. People cut down trees for cultivation and the chimpanzee habitat correspondingly shrinks. The second is poaching and the bushmeat trade. Thanks to the religious beliefs of the Manon people, the chimpanzees at Bossou have been protected, but this is not true in nearby villages where the people are of different ethnic origins. The third threat is contagious diseases. We lost five chimpanzees in Bossou in a flu-like epidemic in 2003 (Matsuzawa, 2006a), and there was a recent tragedy in Mahale, Tanzania where 12 chimpanzees died in 2006, apparently from a disease transmitted by humans (T. Nishida, personal communication).

Figure 17.2 summarizes demographic changes in the Bossou community since 1976 (Sugiyama, 2004; Matsuzawa, 2006b). The maximum number of chimpanzees recorded was 22. The total fell to a minimum of 12 and then recovered to 13 in September 2007. These data show two things. First, while the number of chimpanzees was relatively stable for most of the study period (around 20 individuals), it has fallen drastically in recent years. Second, there are many old chimpanzees and very few young ones in the community, a tendency that has accelerated in recent years. To preserve the Bossou community and its unique culture, we need to expand the chimpanzees' habitat and enable them to exchange genes with a neighboring population.

The "Green Corridor Project" was therefore started in January 1997. The aim of this project was to expand the forests of Bossou to the east in

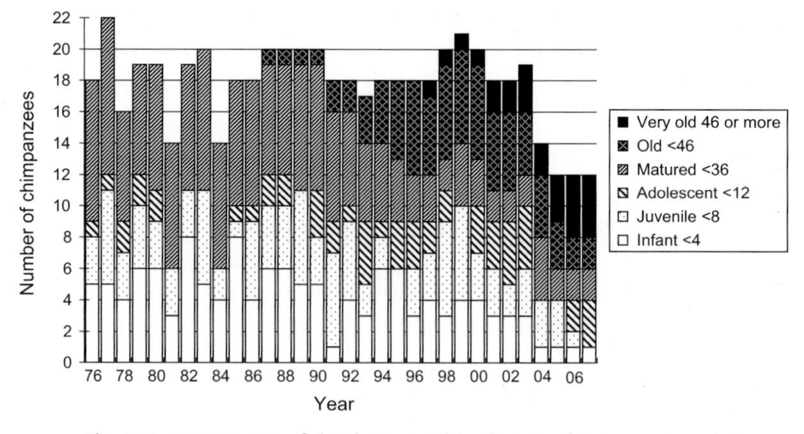

Fig. 17.2. A summary of the demographic change of Bossou community since 1976 (modified from Matsuzawa, 2006b).

order to connect them to those of the Nimba Mountains (Fig. 17.3). In doing so, it was hoped that there would be increased opportunities for the chimpanzees to go back and forth between the two habitats. In October 2001 the government of Guinea helped by establishing an Institute at Bossou, the Institute for Environmental Research at Bossou (IREB), to enable Guinean nationals to collaborate. The Green Corridor Project is a cooperative effort among researchers, governmental employees, and local villagers.

GREEN CORRIDOR PROJECT

The Nimba Mountains are located about 10 km east of Bossou. They are at the borders of three nations: Guinea, Côte d'Ivoire, and Liberia. UNESCO has designated Nimba a World Natural Heritage (WNH) Site in Danger in both Guinea and Côte d'Ivoire.

The author, TM, first reached the summit of Mount Nimba on March 3, 1986. The view from the top clearly showed the strategic importance of this area for chimpanzee conservation in West Africa. Matsuzawa and Yamakoshi (1996) carried out a preliminary survey of the Nimba chimpanzees on the Côte d'Ivoire side in 1993. Thanks to subsequent investigations (Shimada *et al.*, 2004; Humle and Matsuzawa, 2001, 2004; Koops and Matsuzawa, 2006; Koops *et al.*, 2007; Granier, 2007), we know that there is at least one community of chimpanzees in the Seringbara area on the Guinean side of the Nimba Mountains, and that this is the nearest adjacent community of chimpanzees to the Bossou community.

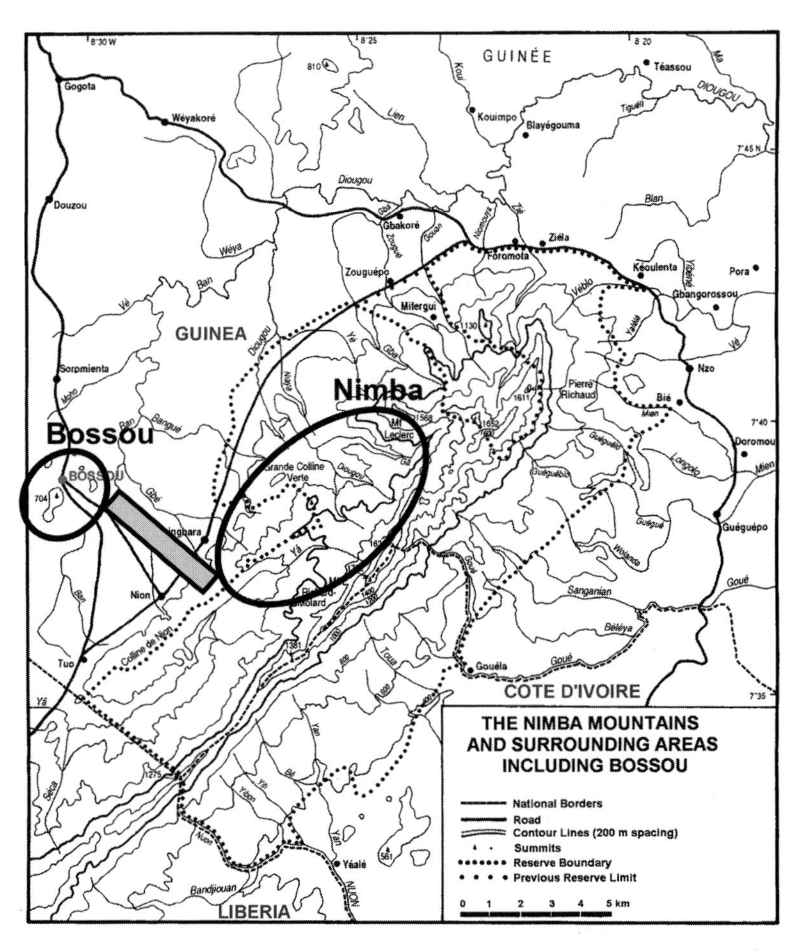

Fig. 17.3. A map of Bossou and Nimba: the Green Corridor Project aimed to connect the two habitats by planting trees in the savanna.

There is savanna stretching for about 4 km between Bossou and Seringbara. There are also small gallery forests along streams. The Green Corridor Project aims to connect Bossou and Nimba by planting trees in the savanna.

According to the old people in Bossou, many years ago the savanna used to be a large forest. So our goal became the restoration of the savanna back to forest. The plan was to plant trees at 5 m intervals in a corridor 4 km long and 300 m wide – an area of 120 ha. To do this, we needed to plant 48 000 trees in total. Based on this plan, we grew 7000–10 000 young trees in a tree nursery each year.

A unique aspect of the plan is that we utilized chimpanzee feces when selecting the tree species to be planted. The aim was to grow trees that produce fruits favored by the chimpanzees. If the purpose were to simply create any forest, we could have chosen tree species that would be more suitable for transplantation such as *Eucalyptus*. Another possibility would have been to plant fruiting trees preferred by people, such as mangoes and oranges. However, we thought that we should focus on making this a forest for chimpanzees by planting trees used by them and not by humans.

THE BOSSOU METHOD OF PLANTING TREES

Thanks to trial and error over the past 10 years, we have successfully established an unusual way of planting trees. It consists of three stages: (a) nursing the young trees, (b) transplanting them into the savanna, and then (c) protecting the growing trees. The details of each stage are as follows.

Nursing young trees

We developed a method of nursing young trees by using chimpanzee feces. Seeds taken from chimpanzee feces come from fruits that they have eaten and will therefore, in turn, populate our new forest with fruiting trees favored by the chimpanzees. Moreover, the seeds that have passed through a chimpanzee's gut have better germination rates than ones simply dropped on the ground (H. Takemoto, unpublished data). The young trees were grown in plastic sacs in the tree nursery. We sometimes also collected small saplings that we found in the forest, and brought them back to the tree nursery. Researchers often encountered clusters of saplings that had originated from chimpanzee feces. Chimpanzees are a key species for seed dispersal and for the maintenance of diversity in tropical rainforests. In a sense, the Green Corridor Project is an attempt to artificially enhance the seed dispersal process by chimpanzees.

Transplanting into the savanna

In the next stage we transplanted the young trees into the savanna, planting them in cultivated ground at 5 m intervals. We also tried a second tactic, growing trees in abandoned fields. We asked the village farmers to plant cassava (*Manihot esculenta*) or rice (*Oryza* spp.) in newly cultivated fields in the savanna. We also asked them to plant young trees between their crops. Once the crops were harvested, we asked them to abandon the fields. This combination of cassava/rice and young trees motivated the farmers to cultivate the savanna. The concept of "forestation" is a little hard for the local people to accept, because they are used to cutting down trees without attempting to replant them. Therefore, a project that combined tree planting with cultivating crops proved to be a good and practical solution.

Protecting young trees

Third, we made an effort to protect the trees that were growing. There are three main reasons why it is difficult to grow trees in the savanna. First, the soil is poor. For example, there were several places in the savanna where an iron-like soil covered the surface of the ground, and no plants grew. Second, the saplings are eaten by cattle and insects. Domestic animals such as sheep, goats, and cows often intrude into the savanna to eat the grass. Third, there are problems caused by humans, such as fire. People set fires in the savanna to burn off old vegetation and encourage the growth of new grass to feed their cattle. They also start fires while hunting agoutis (*Aulacodes* spp.), which live in underground tunnels. To protect the young trees therefore, we set up fire-breaks 10 m wide on both sides of the corridor. We also employed two local people to patrol the fire-break area and to control the cattle. To protect the seedlings, we use "*hexatubes*," as explained below.

ASSESSMENT OF PLANTING ACTIVITY

The first stage of the Green Corridor Project was to create a small botanical garden (Projet Petit Jardin) as a pilot project. The garden was constructed on 0.36 ha (about 60 m × 60 m) in the savanna area on the periphery of the chimpanzee habitat. Several local assistants cut the bush then planted 250 nursery trees from 28 different species, all of which grow in the core area of the chimpanzee habitat at Bossou (Sugiyama and Koman, 1987, 1992). Eighteen months later in July 1998, the trees in the garden were inspected by Hirata and Morimura (Hirata *et al.*, 1998). Fifty percent (125) of the planted trees were still alive. No further planting was done and 8 years later, in 2005, a second inspection was carried out (Matsuzawa, 2007).

For this assessment, we cut the grass and identified all the trees remaining in our original pilot plot. Of the trees that we had planted, 62 (24.8%) had survived in the savanna. Among them were the following four

Fig. 17.4. The savanna was transformed to a secondary forest. You can still recognize the planted *Uapaca heudelotii* trees at 5 m intervals. (Photo by T. Matsuzawa.)

species: *Uapaca heudelotii, Parkia bicolor, Craterispermun laurinum*, and *Albizia zygia*. The tallest tree was a *Parkia bicolor*, which was 9.2 m high. The second tallest was a *Uapaca heudelotii*, which was almost as tall (Fig. 17.4).

In addition to these trees we found 386 other young trees that had not been planted by us. These trees had grown naturally, from seeds brought in by the wind, mammals, or birds. Thus, of the 448 trees growing in this plot, 86.2% had self-seeded during the 8 years. Among the naturally grown trees, we identified 30 species, which is a wide variety in such a small area. Among them, the following three species were dominant: *Harungana madagascariensis* ($n = 55$), *Nauclea latifolia* ($n = 55$) and *Dichrostachys glomerata* ($n = 40$).

From our efforts in the past 8 years, we can draw the following conclusions. First, some of the forest trees, such as *Uapaca heudelotii* and *Parkia bicolor*, can survive when planted in the savanna area. In future plantings we should choose tree species that provide food for the chimpanzees and that survive in the savanna. Second, many naturally seeded trees grew in the savanna area without our planting them. This suggests that guarding the area from browsing animals and preventing bush fires were more important than planting for the success of reforestation programs. Third, we found that we can transform the savanna into forest through reforestation. Based on the initial attempt in the Petit Jardin we

Fig. 17.5. A *hexatube* is a hexagon tube, 1.4 m high, made from polypropylene. The tubes were set to protect the young trees in the savanna. (Photo by G. Ohashi.)

estimate that, in 8 years, we can grow trees reaching close to 10 m in height, using our method of tree nursing followed by planting the young trees in the savanna.

Thanks to the Green Corridor Project, the forests of Bossou have expanded to the east step by step. Apparently as a result, chimpanzees are now often using the reforested area closest to the Nimba Mountains. The number of observations of chimpanzees in the Seringbara area has significantly increased in recent years. They sometimes stay for a few days or even for a week in the forest of Seringbara.

HEXATUBES: A NEW WAY TO PROTECT THE YOUNG TREES

In this final section, we would like to describe our recent use of "*hexatubes*" to protect the young trees. Young trees were planted in the savanna inside 1.4 m high hexagonal polypropylene tubes. The tubes maintain an optimal microclimate for the growing trees by controlling temperature and humidity, and by protecting them from strong winds and grazing (Fig. 17.5).

We first brought 1200 tubes from Japan in September 2005. All of them were in use by December 2005. We shipped an additional 5000 tubes in 2006 and 3000 more in 2007. About 70% of the 3000 young trees grown in the tubes survived their first year. The tubes were also set up in the courtyard of a school to help pupils observe the growth of young trees.

This illustrates a second function of the reforestation program: environmental education. Reforestation is not the only conservation goal of the Bossou project. We also work to change the local people's attitude to their natural environment. Our programs are based on respect for the Manon people, and for the Manon culture which has protected these chimpanzees and their forest for generations.

SUMMARY

The Bossou chimpanzee community, a small and isolated population with a unique culture of tool use, has experienced an alarming drop in size during the last few years. While all young females leave the group, there have been no new chimpanzees immigrating in during the time of the study. In an attempt to link the Bossou chimpanzees with other communities, a novel attempt has been made to reforest the savanna between the home range of the Bossou chimpanzees and the neighboring Nimba Mountain forest. The Green Corridor Project has shown that it is possible to reforest savanna, and suggests that the Bossou community will shortly be in demographic contact with a larger population. This innovative conservation project was a direct result of our long-term research program at Bossou.

ACKNOWLEDGMENTS

The Green Corridor Project was made possible by the collaboration of researchers (KUPRI International team), a governmental institute (IREB), and the local people of Bossou and Seringbara. The long-term project of Bossou and Nimba was supported financially by MEXT-16002001, JSPS-HOPE, JSPS-GCOE-A06 (Biodiversity) to TM. The tree-planting project is partly supported by a grant from the Ministry of the Environment to Dr. Toshisada Nishida. The long-term project has been supported by the following organizations: DNRST and Japanese Embassy in Guinea, Japan Fund for Global Environment, The Japan Trust for Global Environment, Toyota, Phytoculture Control Co Ltd, USFWS, CI (PAF), and Houston Zoo. Special thanks are due to Dr. Kabine Kante, Dr. Tamba Tagbino, Mr. Ryo Hasegawa, Mr. Gaku Ohashi, Mr. Paquile Cherif, Mr. Gouano Goumi, Mr. Tino Zogbira, Mr. Nicolas Granier, and Dr. Tatyana Humle for their help for the Green Corridor Project at each stage. The details of the long-term project are introduced on the following website: http://www.greenpassage.org/indexE.rhtml.

REFERENCES

Emery Thompson, M., Jones, J. H., Pusey, A. E. *et al.* (2007). Aging and fertility patterns in wild chimpanzees provide insights into the evolution of menopause. *Current Biology*, **17**, 1–7.

Granier, N., Huynen, M. C., and Matsuzawa, T. (2007). Preliminary surveys of chimpanzees in Gouéla area and Déré Forest. *Pan Africa News*, **14**, 20–22.

Hirata, S., Morimura, N., and Matsuzawa, T. (1998). Green passage plan (tree-planting project) and environmental education using documentary videos at Bossou: a progress report. *Pan Africa News*, **5**, 18–20.

Hockings, K. J., Anderson, J. R., and Matsuzawa, T. (2006). Road crossing in chimpanzees: a risky business. *Current Biology*, **16**, R668–R670.

Hockings, K. J., Humle, T., Anderson, J. R. *et al.* (2007). Chimpanzees share forbidden fruit. PLoS ONE 2, e886. doi:10.1371/journal.pone.0000886.

Humle, T. and Matsuzawa, T. (2001). Behavioural diversity among the wild chimpanzee populations of Bossou and neighbouring areas, Guinea and Côte d'Ivoire, West Africa – a preliminary report. *Folia Primatologica*, **72**, 57–68.

Humle, T. and Matsuzawa, T. (2004). Oil palm use by adjacent communities of chimpanzees at Bossou and Nimba Mountains, West Africa. *International Journal of Primatology*, **25**, 551–581.

Koops, K. and Matsuzawa, T. (2006). Hand clapping by a chimpanzee in the Nimba Mountains, Guinea, West Africa. *Pan Africa News*, **13**, 19–21.

Koops, K., Humle, T., Sterck, E. H. M., and Matsuzawa, T. (2007). Ground-nesting by the chimpanzees of the Nimba Mountains, Guinea: environmentally or socially determined? *American Journal of Primatology*, **69**, 407–419.

Kortlandt, A. (1986). The use of stone tools by wild-living chimpanzees and earliest hominids. *Journal of Human Evolution*, **15**, 77–132.

Lamotte, M. (1942). La faune mammalogique du Mont Nimba (Haute Guinée). *Mammalia*, **6**, 114–119.

Matsuzawa, T. (1994). Field experiments on use of stone tools in the wild. In *Chimpanzee Cultures*, ed. R. W. Wrangham, W. C. McGrew, F. B. M. de Waal, and P.G. Heltne. Cambridge: Harvard University Press, pp. 351–370.

Matsuzawa, T. (2006a). Bossou 30 ans. *Pan Africa News*, **13**, 16–19.

Matsuzawa, T. (2006b). Sociocognitive development in chimpanzees: a synthesis of laboratory work and fieldwork. In *Cognitive Development in Chimpanzees*, ed. T. Matsuzawa, M. Tomonaga, and M. Tanaka, Tokyo: Springer, pp. 3–33.

Matsuzawa, T. (2007). Assessment of the planted trees in the Green Corridor Project. *Pan Africa News*, **14**, 27–29.

Matsuzawa, T. and Yamakoshi, G. (1996). Comparison of chimpanzee material culture between Bossou and Nimba, West Africa. In *Reaching into Thought: The Mind of the Great Apes*, ed. A. E. Russon, K. A. Bard, and S. Parker. Cambridge: Cambridge University Press, pp. 211–232.

McGrew, W. (2004). *The Cultured Chimpanzee: Reflections on Cultural Primatology*. Cambridge: Cambridge University Press.

Ohashi, G. (2006). Behavioral repertoire of tool use in the wild chimpanzees at Bossou. In *Cognitive Development in Chimpanzees*, ed. T. Matsuzawa, M. Tomonaga, and M. Tanaka. Tokyo: Springer, pp. 439–451.

Ohashi, G. (2007). Papaya fruit sharing in wild chimpanzees at Bossou, Guinea. *Pan Africa News*, **14**, 14–16.

Shimada, M. K., Hayakawa, S., Humle, T. *et al.* (2004). Mitochondrial DNA genealogy of chimpanzees in the Nimba Mountains and Bossou, West Africa. *American Journal of Primatology*, **64**, 261–275.

Sugiyama, Y. (2004). Demographic parameters and life history of chimpanzees at Bossou, Guinea. *American Journal of Physical Anthropology*, **124**, 154–165.

Sugiyama, Y. and Koman, J. (1979). Social structure and dynamics of wild chimpanzees at Bossou, Guinea. *Primates*, **20**, 323–339.

Sugiyama, Y. and Koman, J. (1987). A preliminary list of chimpanzees' alimentation at Bossou, Guinea. *Primates*, **28**, 133–147.

Sugiyama, Y. and Koman, J. (1992). The flora of Bossou: its utilization by chimpanzees and humans. *African Study Monographs*, **13**, 127–169.

Yamakoshi, G. (2001). Ecology of tool use in wild chimpanzees: toward reconstruction of early hominid evolution. In *Primate Origins of Human Cognition and Behavior*, ed. T. Matsuzawa, Tokyo: Springer. pp. 537–556.

ELIZABETH A. WILLIAMSON AND KATIE A. FAWCETT

18

Long-term research and conservation of the Virunga mountain gorillas

BACKGROUND

The Virunga Volcanoes encompass three National Parks in three countries of eastern Central Africa: Mgahinga Gorilla National Park in Uganda, Virunga National Park in the Democratic Republic of the Congo and Volcanoes National Park in Rwanda (Fig. 18.1). This region harbors one of only two remaining populations of mountain gorillas, 380 "Virunga" gorillas (*Gorilla beringei beringei*). The Virungas cover an area of about 425 km^2 and contain a variety of afromontane habitats, stratified by altitude ranging from 1850 m to 4507 m above sea level. Much of this high altitude vegetation is not suitable for the gorillas (Weber and Vedder, 1983), thus the gorilla population is concentrated below 3400 m in the mid-altitude *Hagenia–Hypericum* zone and the lower altitude bamboo zone.

The first National Park in Africa was created in 1925, specifically to protect the mountain gorillas. These magnificent beasts received little attention until 1959, by which time they were thought to number only 400–500 individuals (Schaller, 1963). Following a pioneering study by George Schaller, long-term research and conservation efforts began in 1967 when Dian Fossey established the Karisoke Research Center in Rwanda. Fossey's study was initiated along the same lines as Jane Goodall's research on chimpanzees at Gombe in Tanzania, after a meeting with the famous paleoanthropologist, Dr. Louis Leakey.

By the 1970s, the Volcanoes National Park had been reduced to 46% of its original size, so that only 160 km^2 of forest remained in Rwanda. The bulk of this habitat conversion was for a pyrethrum project, which excised 100 km^2 of forest in 1968. All forest between 1600 and 2600 m ASL was removed, and an estimated 40%–50% decline in the number of

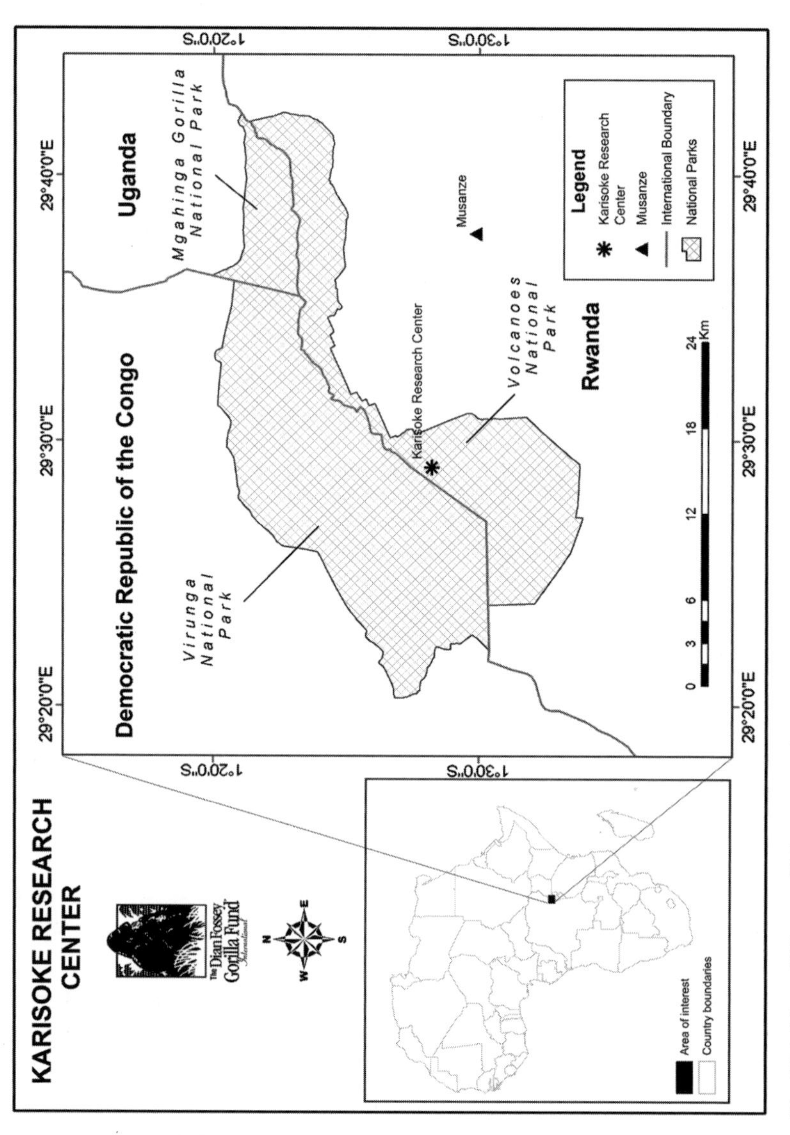

Fig. 18.1. Map of Virunga Volcanoes region.

gorillas ensued (Weber and Vedder, 1983). The population stabilized with the implementation of measures against poachers and their snares in the 1980s and, with increased protection, more gorillas survived in the groups that were monitored daily, and the population increased for the first time in three decades. Today, despite a significant recovery, only a few hundred of these great apes exist and they are classified as Critically Endangered on the IUCN Red List of Threatened Species.

Security has been an overriding factor impacting conservation and research, as the Great Lakes region has been severely affected by fighting between rebel groups and regular armies. More than a decade of civil war and political instability have impacted the Virungas, at times threatening the lives of both gorillas and field staff, and preventing all field operations. Refugees and militia invaded the forest periodically between 1991 and 2001, causing severe habitat degradation by cultivating food in the Park, as well as increasing the likelihood of disease outbreaks, with many people living in unhygienic and unhealthy conditions (Plumptre and Williamson, 2001).

Since 1959, research has made important contributions to mountain gorilla conservation, and we present examples in the following sections.

ECOLOGY AND BEHAVIOR OF MOUNTAIN GORILLAS

As with chimpanzees, mountain gorillas must be "habituated" before details of their social behavior can be studied. Habituation is a process by which wild animals become accustomed to the presence of humans in their vicinity to the point where humans are considered a neutral element in their environment. This requires considerable investment of time and effort, months or more often years, and the key to success is persistent, regular, and frequent neutral contact with the same individuals (Williamson and Feistner, 2003). Anyone considering habituation for research or tourism purposes should bear in mind that habituated individuals are more vulnerable in dangerous situations, since habituation is basically the loss of fear of humans. Habituation makes animals easier to approach, not only by Park staff but also by militias and poachers. The future of any study or tourism population must be contemplated carefully before making the decision to habituate, as their lifelong protection must then be assured.

Habituation allows observers to approach within a few meters, because subjects no longer flee, so behavior can be sampled consistently and fine level behaviors such as subtle social interactions or food processing can be observed. Observers can also learn to identify individuals, which is essential for research purposes. In the habituated mountain gorilla groups, each animal is named, photographed, its "noseprint" drawn, other physically distinctive characteristics noted, and demographic information recorded systematically. An individual's mother, date of birth, sex, and other pertinent information are noted as a matter of course. Daily observations of gorillas allow field staff to check on the physical condition of each individual, to remove snares from their habitat, and to intervene rapidly in emergency situations. Precise numbers of habituated gorillas in the Virungas are known through this close monitoring of identified individuals.

A small population of slow-reproducing and long-lived great apes must be studied for many years before we can begin to establish "norms" for their development and behavior. Like the study of chimpanzees in Kibale National Park, Uganda, the long-term research at Karisoke is one of a handful of continuous studies of an animal population that has spanned decades. Conservation practitioners are able to use the results of this research to interpret dramatic instances of natural behavior, such as infanticide, while knowledge of the natural processes of male emigration and female immigration explains transfers and "disappearances" of individuals (e.g., Sicotte, 1993). Data on interbirth intervals and other reproductive parameters are critical to assessing rates of change (Harcourt *et al.*, 1981). Demographic and life history data are especially important for Protected Area managers to evaluate the effectiveness of their activities. A steady increase in mountain gorilla numbers in the 1980s indicated that conservation actions were having positive results.

Similarly, an understanding of gorilla feeding ecology, nutrition, and ranging behavior is important in determining whether the Virunga population could increase in size by occupying more of the habitat, and for improving management practices in areas which are not used by gorillas. For example, ecological data will be crucial to evaluate any "underuse" of parts of the forest (Watts, 1998). If able to assess avoidance of certain areas, Protected Area managers may be able to alleviate detrimental conditions and improve the gorillas' chances of survival.

Tourism is considered below, but we should mention here that the research center played a critical role in developing the techniques of gorilla habituation: researchers with experience gained at Karisoke designed and initiated gorilla tourism in Rwanda. This successful program incorporated knowledge of the research groups' diet, daily travel distance and ranging, to anticipate group movements, and locate the gorillas with relative ease. Predictability of daily activity rhythms was also important and, when possible, visits were timed to coincide with the gorillas' rest periods, facilitating observation conditions for the visitors (Plumptre and Williamson, 2001).

Another concern for the future is whether the size of the mountain gorillas gene pool has been reduced to a level where inbreeding may become a serious problem. A Population and Habitat Viability Assessment (PHVA) and similar analyses have shown that habitat loss is a greater danger to gorilla survival than inbreeding (Harcourt, 1995). Such studies can guide the use of resources and the development of Park management plans.

Many generalizations and assumptions about gorillas have been made, based on our knowledge of only one small population in Rwanda. Western gorillas are much more difficult to study; they rarely lose their fear of humans because:

(a) the vegetation and the terrain of their flat dense forest habitat make observation difficult, and the gorillas cannot see the trackers from a safe distance;
(b) their lowland habitat is quite different from montane forest, consequently their ecology is different. Their ranging behavior, in particular, makes them difficult to follow as they leave little trail where they have been feeding, and do not make trails in the vegetation;
(c) they are often hunted as bushmeat;
(d) they have sometimes been hunted for "sport".

Consequently, the mountain gorilla model has been utilized not only by other projects in the Great Lakes region, but in the absence of data on western gorillas, research carried out in Rwanda has also provided valuable input to gorilla research and tourism programs in Cameroon, Central African Republic, Gabon and the Republic of the Congo. Similarly, the chimpanzee populations of the Congo Basin have barely been studied and so conservation programs are designed using knowledge of the ecological and behavioral needs of chimpanzees in Tanzania and Uganda.

TOURISM

International awareness of the research center opened the door to tourism. Dian Fossey and now her legacy have been a major draw; even the remains of Karisoke form part of the tourism circuit. Although strictly speaking neither research nor conservation, links between tourism and research are significant, and will be discussed here.

In 1979 plans were announced to convert a large swathe of Virunga Parkland to cattle pasture. At that time, habitat destruction was the greatest threat to gorilla survival and so a means of maintaining the forest and making the gorillas "pay for themselves" was needed urgently. A highly regulated tourism program was initiated, founded on experience gained at Karisoke. Visits by tourists are potentially stressful to the gorillas, so it was important to minimize risks to both gorillas and people. Consequently, important rules were established regarding the number of visitors permissible, the distance to which gorillas can be approached, and a 1-hour time limit imposed. Gorilla tourism gradually became a great success in terms of much-needed revenue, which provided for increased protection of the habitat, and close surveillance of additional gorilla groups. In 1989, tourism was the third highest foreign currency earner for Rwanda after tea and coffee (Weber, 1993), and the revenue earned directly from gorilla tourism is now estimated at US$3 million per annum. However, financial benefits extend beyond the price of a permit, and tourists brought US$35 million to Rwanda in 2006 (*New York Times*, 2007).

Tourism with great apes has well-known pros and cons (see also Mugisha, Chapter 11). The greatest risk to gorillas is the potential for introduction of lethal diseases from humans – guides and trackers as well as tourists. With such a small number of gorillas remaining, an infectious disease could devastate the population. While this was recognized at the start of the tourism program, loss of their habitat was a far greater threat to the gorillas at that time. Strict rules were put in place for gorilla tourism, and adherence to these rules is vital to minimizing the associated risks.

Until recently, we relied on speculation, extrapolation, and common sense to evaluate the risks of cross-infection between gorillas and humans. Now the tourism regulations have been reviewed in the light of epidemiological data: studies of captive gorillas show that they are susceptible to human diseases, but do not have the same defences as humans (Homsy, 1999). While most international tourists visiting Rwanda are fairly healthy, and have been inoculated against certain diseases, many pick up respiratory infections on long-haul flights. Illnesses to which the gorillas have never been exposed are potentially the most dangerous.

Soon after tourism began, the Volcanoes Veterinary Center was established in direct response to need – Dian Fossey had determined that declining gorilla numbers in the 1980s were due to human-caused disease and injuries. The veterinary project developed a health-monitoring program for gorillas, and in recent years has expanded this to include park staff and researchers, providing medical counseling and treatment of common infections, such as intestinal parasites.

Twenty years after tourism was initiated, scientific studies of the impacts of tourism on the gorillas have begun. Targeted research is needed to evaluate impacts, both positive and negative, and to provide information to ensure that tourism is implemented sustainably. Continued study of gorillas in the research groups also provides a baseline from which to judge the impacts of tourism. These data allow conservation practitioners to assess whether new or altered behaviors observed might result from stress caused by tourism (Plumptre and Williamson, 2001).

An important means of coping with tourism demands and assuring adequate revenue for the Protected Area authorities is to revise the price of permits regularly. The mountain gorilla population is too small and too fragile to withstand increasing pressure from tourism. Even with high permit fees, currently US$500 per person, numbers of visitors do not diminish, but some of the pressures subside, while the revenue accrued by the governing authorities is maintained or increased.

Some of the major lessons learned in relation to tourism have been the following:

(a) Tourism with such a vulnerable species requires strict enforcement of rules. All sites which promote tourism with great apes should have peer-reviewed guidelines with limits to the duration of visits, the number of visitors permitted, and distance to be maintained between people and gorillas. Projects should include the following components: staff training, health monitoring (apes and people), and guidance in appropriate visitor control and behavior.
(b) Despite the dangers inherent in tourism, it provides a mechanism for ensuring that great apes and their habitats are valued for many reasons. Tourism has probably saved the gorillas in Rwanda from further habitat loss or degradation.
(c) The high cost of permits is a necessary means to try to limit the pressures put upon both the apes and the Park authorities.
(d) Research and tourism activities should be separate. One important aspect of tourism with mountain gorillas is that tourists do not visit the research groups and do not therefore disrupt research, but this is not always the case elsewhere (e.g., Gombe).
(e) We do not advocate habituation of western gorillas for tourism. It is often assumed that the success seen with mountain gorillas could be exported to the Congo Basin, but this is not the case, in part for the reasons listed above.

(f) Finally, it should never be forgotten that tourism with gorillas was started, first and foremost, as a means of conserving the gorillas, and not simply for financial gain.

LONG-TERM MONITORING OF THE VIRUNGA GORILLAS

Routine censuses to monitor changes in a population are essential to the understanding of population dynamics, and for conservation practitioners to assess the effectiveness of management strategies. Census results and population statistics not only show changes in the actual numbers of gorillas, but also reproductive health and potential growth are indicated by the age–sex composition of the population (Weber and Vedder, 1983).

In 1959, George Schaller conducted the first extensive study of mountain gorillas and he developed a census technique using nest counts and measurements of dung diameter to estimate population size (Schaller, 1963). Censuses have been carried out at more-or-less 5-year intervals since research on mountain gorillas first began. Surveys in the 1970s showed a drastic decrease from 400–500 to only 250 animals. Immediate protective measures were imposed to prevent further decline. These measures, described below, have led to a slow recovery and growth of the population (Fig. 18.2).

One aspect of long-term monitoring, lacking until relatively recently, was any detailed monitoring of groups outside of the area covered by the research center. Ranger-based monitoring has been developed and implemented as a data collection tool for Park managers (Gray and Kalpers, 2005). It is simple and systematic, and has been applied to all gorilla groups habituated for tourism, greatly expanding our overall knowledge of gorilla demography and habitat use throughout the Virungas.

Population monitoring has also enabled us to assess the effects of war and instability from 1991 and 2001. It is notable that increases in the Virunga population can be accounted for by one subsection of the population: the Karisoke research groups plus the Susa tourist group. If other groups have not fared so well, this is likely to result from differing levels of protection, human disturbance, and demographic factors (Kalpers *et al.*, 2003).

Population modeling has shown that the Virunga population is viable for at least the next 100 years in the absence of severe disturbance, but that the population could easily suffer a heavy decline in the event of environmental perturbations such as habitat loss or degradation. A malicious new threat has emerged in the DRC: at least eight gorillas were shot dead in three incidents in 2007. News reports indicated that these deliberate killings were intended to deter conservation activities, thus facilitating the lucrative but illegal production of charcoal inside the Virunga National Park (BBC, 2007).

SURVEILLANCE OF ILLEGAL ACTIVITIES

Anti-poaching patrols, which stemmed from the research program, are an essential aspect of Park management and biodiversity protection. To combat the illegal killing of antelope and buffalo for their meat, Dian Fossey initially tried to thwart poacher activities by cutting their trap-lines. She also herded gorillas away from areas where snares had been set, as they could become unintended victims. But, after the slaying of several gorillas in attempts to capture their infants, Fossey employed anti-poaching teams and established regular patrols in 1978. The formation of the Mountain Gorilla Project a year later increased and improved patrols and law enforcement, and thus discouraged interest in gorilla infants. The subsequent creation of the regional International Gorilla Conservation Program led to greater collaboration, coordination of ranger patrols and sharing of information among the three range states. In recent years, the gorilla groups that have been monitored most closely for either research or tourism have experienced the highest growth rates (Kalpers *et al.*, 2003).

Research into poaching has included an assessment of the frequency and location of snares in the park to determine patrol effectiveness. In the 1990s, an analysis of patrol effort showed that, as the number of patrols increased, more snares were found, but that the number of snares per patrol dropped after patrols reached 20 days per month. There were peaks in poaching around Christmas and Easter, when households need extra disposable income, which necessitated extra protection efforts at these times (Plumptre and Williamson, 2001).

During the last decade, ranger-based monitoring (RBM) has become a key management tool throughout the Virungas. Information is collected not only on the gorillas, but also on illegal use of resources in the forest. Knowledge of the distribution of illegal activities allows patrol coverage to be targeted. The level of illegal activities is now high at all times of year, reflecting human demands for natural resources. RBM data have shown that, in recent years, increased patrol effort has not necessarily resulted in a reduction in illegal activities, and that it is also necessary to investigate the links between resource utilization and the economic situation of local people. The gorillas' habitat is surrounded

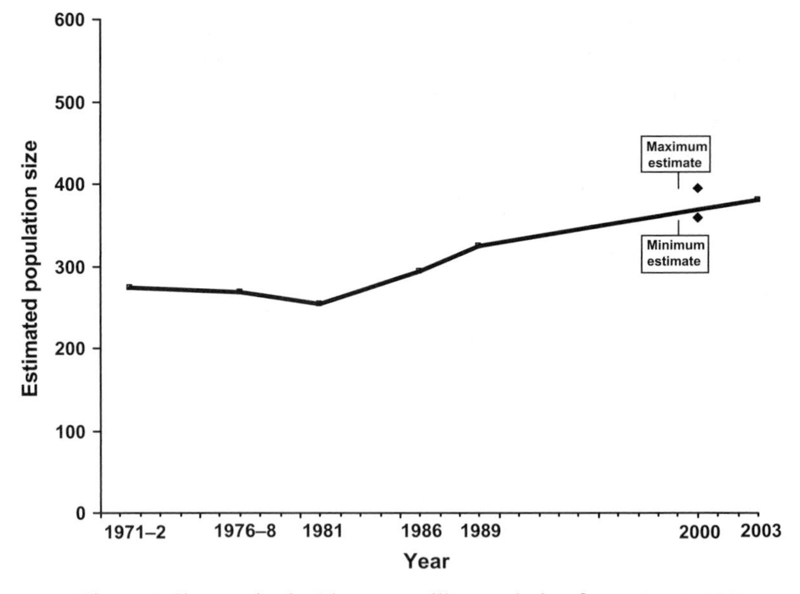

Fig. 18.2. Changes in the Virunga gorilla population from 1971 to 2003.

by one of the highest human population densities in Africa, at 400–600 people per km^2. The fertile volcanic soil supports this high-density human population of subsistence farmers, and there is constant pressure along the Park edges for additional agricultural land. Activities are now being developed to assist local communities to meet their needs, with the intention of reducing pressure on the forest (Gray and Kalpers, 2005).

As a result of renewed direct poaching of gorillas since 2002, there are currently 11 confiscated infants (four mountain and 7 Grauer's gorillas) in the care of the Mountain Gorilla Veterinary Project, Dian Fossey Gorilla Fund International, and the Rwandan and Congolese National Park authorities. This partnership of veterinary and behavioral specialists provides the care necessary for these orphans to recover from the devastating trauma of poaching. To date, no mountain gorillas have been successfully reintroduced to the wild (Whittier and Fawcett, 2006). It is proposed that, by caring for and rehabilitating a young female mountain gorilla, Maisha, until she is age-appropriate for a transfer between gorilla groups in the wild (i.e., adolescent), it may be possible to reintroduce her to a wild mountain gorilla group or lone silverback. Behavioral data from Karisoke are guiding decisions about diet, socialization, and the future choice and timing of any reintroduction attempts. This plan, if successful, will serve as a model for mountain gorilla reintroductions.

Fig. 18.3. A silverback male mountain gorilla and family. © David Pluth.

ECOLOGICAL RESEARCH IN THE VIRUNGAS

The Virungas form part of the Albertine Rift biodiversity hotspot, harboring an exceptionally high number of endemic plants and animals, and the scope of research activities has broadened over time to include other aspects of the Park's biodiversity. Recently, research has expanded to cover key species and habitats, focusing on endemics such as the Golden Monkey (*Cercopithecus mitis kandti*) and Grauer's swamp warbler (*Bradypterus graueri*).

An ecosystem approach, in addition to a focus on gorillas, has provided valuable information to bolster the arguments for keeping the forest cover on the volcanoes. In the 1980s, it was demonstrated that converting the entire Park to agricultural land would only provide extremely marginal land for the equivalent of 1 year of Rwanda's human population growth. The slopes of the volcanoes are too steep for sustainable agricultural production and could only provide short-term benefits rather than the longer-term benefits of gorilla tourism. Regional land use could not increase in any significant way in the Ruhengeri prefecture (Weber, 1987).

A high level of rainfall is generated in the volcanoes, and this rain is captured by the forest in a way that minimizes erosion and provides a gradual release of water into surrounding areas, and thus a perennial supply of clean water. The Volcanoes National Park forms only 0.5% of

Rwanda's surface area but contributes a vital 10% of the water catchment (Weber, 1987).

GIS and remote sensing have become established technologies in conservation management. Satellite imagery has been used to produce vegetation maps of the Virungas to address questions about the habitat's carrying capacity, the impacts of illegal human activities, and changes in the type and extent of vegetation cover over time.

Monitoring is essential to effective management. Ecological functions of the forest must be demonstrated, tourism and other impacts on the ecosystem should be monitored, to enable us to develop and improve tools with which to protect the Park and the gorillas. The positive integration of research and management will ensure that management decisions are based on sound scientific data.

CAPACITY BUILDING

A crucial role of the Karisoke Research Center is to contribute to the scientific training of the next generation of conservation biologists. Rwanda has the political will to protect mountain gorillas and their habitats; however, much of the contemporary science has been conducted by expatriate researchers. It is important to build upon the ability of Rwandan scientists and conservation managers to conduct conservation-oriented research and implement their findings. Reports, theses, and publications have been written, but the human knowledge base developed during the research process is not always readily available in-country to assist with the implementation of conservation strategies. Efforts must be made to rectify this.

Courses at the National University were interrupted by the genocide in 1994, resulting in a lack of national scientists to lead conservation efforts, a situation common to technical capacities of all sectors in Rwanda. Post-genocide, the Government's strategic plan relies upon the development of human and institutional resources. Given its 40-year history, the Karisoke Research Center can play an important role in supporting the education of Rwandan scientists. Currently, around 150 students participate annually in field-based conservation courses at the research center, while others conduct research for their undergraduate dissertations. This provides opportunities to train young scientists for the future, and also to address important Park management needs: student projects address priority areas of research, identified in Park management plans, and are often of an applied nature (e.g., crop-raiding, tourism values, vegetation dynamics).

SOCIOECONOMIC RESEARCH

We indicated earlier that illegal activities within the Park are a concern for the protection of mountain gorillas. These activities are intended mainly to meet the subsistence needs of the poorest people around the Park. Like many National Parks in Africa, the Volcanoes National Park was managed initially following a strict exclusionary Protected Area model. However, a key focus of contemporary conservation strategies is on local communities living around the Park, with the intention of addressing local welfare needs and mitigating some of the conservation threats caused by human poverty.

The socioeconomic challenges facing conservation in the Virungas are land shortages, high human population growth, high human mortality, immigration of young men, low levels of literacy, and extreme poverty. The consequences have been high rates of soil erosion and loss of soil fertility, flooding, deforestation, and loss of biodiversity through habitat destruction and hunting, problems exacerbated during the 1997–1998 insurgency in the northwest of Rwanda.

The presence of subsistence farming households around the Park has serious implications for conservation. Many households close to the Park depend on resources in the forest, such as clean water, bamboo, grass for thatching, honey, medicinal plants, and bushmeat. In addition, these households face the difficulties of inadequate farmland, little prospect of employment, and no access to credit, and therefore possess few livelihood alternatives. Moreover, the negative aspects of living next to a Protected Area, such as crop-raiding by wild animals, hit the poorest families hardest and reinforce negative attitudes towards the Park (Plumptre *et al.*, 2004). Thus local people meeting subsistence needs may be considered as one of the greatest threats to survival of the mountain gorilla and the integrity of their habitat (Bush, 2004).

Addressing social threats to conservation through local community development initiatives is seen currently as a key method of mitigating such problems. Integrated Conservation and Development Projects (ICDPs) are now a common approach in developing countries. It is usually assumed that ICDPs result in the conservation of natural resources, while at the same time benefiting local communities who may forego less environmentally friendly activities. However, quantitative assessments of ICDP strategies are rare and have shown variable results in terms of meeting either environmental or welfare goals. A key challenge to the successful implementation of social and economic development programs intended to meet conservation objectives is a rigorous understanding of the dynamics of use of the Park by local people in order to design successful interventions. However, conservationists whose core skills lie mainly in the biological sciences may not be best equipped to identify and implement development programs; conservation practitioners therefore must work closely with development organizations to ensure the survival of these forests.

Given that the principal sources of threat to the gorillas and their habitat are local people and their social and economic circumstances, socioeconomic data are needed to qualify and quantify these threats. Are community conservation approaches improving conservation prospects? To date, socioeconomic research has been largely qualitative, focusing on identifying the means by which people live, and on the importance of Protected Area resources in their lives; however, few quantitative data are available on this dynamic. Such data, coupled with relevant biological data from within the Park, could provide an important baseline from which to assess the impact of community conservation projects on human welfare and conservation. Importantly, they can guide the planning of project interventions (Bush, 2004). A study carried out in 2002 showed that people's attitudes towards conservation are improving; however, benefits from tourism are perceived to accrue primarily at a national rather than local level, thus equity issues must be addressed (Plumptre *et al.*, 2004). In addition, a broader valuation of ecosystem services will enable governments to make informed decisions about conservation and management strategies and contribute to the economic justification for financing conservation.

CONSERVATION EDUCATION

An important responsibility of the research community is the dissemination of research results, most commonly through scientific publications. However, in a conservation context it is also important that this information be translated into a form readily accessible to various stakeholders – local, national, and international. The results of research at Karisoke have formed the basis of several education initiatives, which have been successful in raising awareness of the conservation value of the forest and, in particular, of ecosystem services.

SUMMARY

Research on the "Karisoke" gorillas has attracted worldwide attention, fueled through a multitude of nature documentaries and magazine articles about their lives, notably *National Geographic* magazine and the movie "Gorillas in the Mist". Public commitment to try to save the mountain gorillas provides funds for conservation activities through donors and international NGOs. This high profile also attracts many visitors to Rwanda, bringing revenue to the country and reinforcing pride in the national heritage. A positive image enhanced by a well-managed tourism program has stimulated development and generated publicity.

These factors all contribute to strong government commitment to gorilla conservation, which is key to their survival. In a region where the demand for land is intense, the risk of forest conversion for human settlement, cattle pasture, or agriculture has not been eliminated completely. In the last decade, there have been several attempts by local political leaders to degazette areas of the National Park, but intervention by National Government prevented further loss of the forest. Government commitment is also seen in the security provided to researchers and tourists on a daily basis in the form of military escorts. The Congolese portion of the Virungas is already a UNESCO World Heritage Site, and procedures are under way to award this status to the Rwandan sector.

Dian Fossey predicted that mountain gorillas could become extinct within the same century that they were recognized scientifically. Without the attention generated by the research center, support developed though tourism, and consequent commitment from the governments, gorillas would perhaps no longer exist in the Virungas.

The Karisoke Research Center recently celebrated 40 years of almost continuous study of the gorillas. Field activities were suspended periodically during the 1994 genocide and subsequent insecurity, throughout which it was vital to support the staff and their families, and to ensure their safety. To this end, the gorilla trackers and anti-poaching patrols have undergone paramilitary training and have endured difficult and dangerous working conditions. Much of the success of the research and conservation programs, and the continued survival of the mountain gorillas, can be attributed to the long-term dedication and hard work of the field staff of Karisoke and of the Rwandan Office of Tourism and National Parks.

ACKNOWLEDGMENTS

We thank the Rwandan Office of Tourism and National Parks for permission to work in the Volcanoes National Park. We also thank Glenn Bush and Maryke Gray for comments on the manuscript, Eugene Kayijamahe for providing Fig. 18.1, and IGCP for Fig. 18.2. The Karisoke Research Center is funded and operated by the Dian Fossey Gorilla Fund International.

REFERENCES

British Broadcasting Corporation (2007). Concern over mountain gorilla 'executions'. http://news.bbc.co.uk/2/hi/science/nature/6918012.stm (19/9/07).

Bush, G. K. (2004). Conservation management in Rwanda: a review of socioeconomic and ecosystem values. Project Report, Global Environment Facility, Protected Areas and Biodiversity Project. Kigali: Rwandan Environmental Management Authority.

Fossey, D. (1983). *Gorillas in the Mist.* Boston: Houghton Mifflin.

Gray, M. and Kalpers, J. (2005). Ranger based monitoring in the Virunga–Bwindi region of East–Central Africa: a simple data collection tool for park management. *Biodiversity and Conservation*, **14**, 2723–2741.

Harcourt, A. H. (1995). Population viability estimates, theory and practice for a wild gorilla population. *Conservation Biology*, **9**, 134–142.

Harcourt, A. H., Fossey, D., and Sabater Pi, J. (1981). Demography of *Gorilla gorilla*. *Journal of Zoology*, **195**, 215–233.

Homsy, J. (1999). *Ape Tourism and Human Diseases: How Close Should We Get?* Kampala: International Gorilla Conservation Program http://www.igcp.org/files/ourwork/Homsy_rev.pdf.

Kalpers, J., Williamson, E. A., Robbins, M. M. *et al.* (2003). Gorillas in the crossfire: assessment of population dynamics of the Virunga mountain gorillas over the past three decades. *Oryx*, **37**, 326–337.

New York Times (2007). Once ravaged by war, now vacation spots. http://travel.nytimes.com/2007/09/01/business/worldbusiness/01tourism.html 1/9/07.

Plumptre, A. J. and Williamson, E. A. (2001). Conservation oriented research in the Virunga region. In *Mountain Gorillas: Three Decades of Research at Karisoke*, ed. M. M. Robbins, P. Sicotte, and K. J. Stewart. Cambridge: Cambridge University Press, pp. 361–390.

Plumptre, A. J., Kayitare, A., Rainer, H. *et al.* (2004). *The Socio-economic Status of People Living Near Protected Areas in the Central Albertine Rift.* New York: Albertine Rift Technical Reports, **4**, 127.

Schaller, G. B. (1963). *The Mountain Gorilla: Ecology and Behavior.* Chicago: University of Chicago Press.

Sicotte, P. (1993). Inter-group encounters and female transfer in mountain gorillas: influence of group composition on male behavior. *American Journal of Primatology*, **30**, 21–36.

Watts, D. P. (1998). Long-term habitat use by mountain gorillas (*Gorilla gorilla beringei*). Reuse of foraging areas in relation to resource abundance, quality, and depletion. *International Journal of Primatology*, **19**, 681–702.

Weber, A. W. (1987). *Ruhengeri and its Resources: An Environmental Profile of the Ruhengeri Prefecture.* Kigali: ETMA/USAID, 171.

Weber, A. W. (1993). Primate conservation and ecotourism in Africa. In *Perspectives on Biodiversity: Case Studies of Genetic Resource Conservation and Development*, ed. C. S. Potter, J. I. Cohen, and D. Janczewski. Washington DC: AAAS Press, pp. 129–150.

Weber, A. W. and Vedder, A. L. (1983). Population dynamics of the Virunga gorillas 1959–1978. *Biological Conservation*, **26**, 341–366.

Whittier, C. and Fawcett, K. (2006). Application of the RSG Guidelines in the case of confiscated mountain gorillas, Virunga Massif: Rwanda, Uganda and DRC. *Re-introduction News*, **25**, 40–41.

Williamson, E. A. and Feistner, A. T. C. (2003). Habituating primates: Processes, techniques, variables and ethics. In *Field and Laboratory Methods in Primatology: A Practical Guide*, ed. J. M. Setchell and D. J. Curtis. Cambridge: Cambridge University Press, pp. 25–39.

NATARAJAN ISHWARAN

19

Long-term research and conservation of great apes: a global future

INTRODUCTION

This book stems from a workshop that celebrated the twentieth anniversary of the Kibale Chimpanzee Project. Research on chimpanzees in Uganda, together with studies in Gombe and Mahale in Tanzania and elsewhere, has challenged our assumptions of culture being a uniquely human trait, if by culture one means any widespread behavior that is transmitted by learning rather than acquired by inheritance. In addition to chimpanzees *Pan troglodytes*, the other great apes (bonobos *Pan paniscus*, gorillas *Gorilla gorilla* and orangutans *Pongo pygmaeus*), as well as elephants, whales, and various kinds of birds, show evidence of culture in the wild (Fernandez-Armesto, 2004). Together with their close genetic relationship to humans, this is an important reason why researchers study great ape ecology and behavior, and why many people hope to conserve these charismatic species.

The relationship between field research and scientific conclusions has been a mutually beneficial one; but this has not always been true for the relationship between research and conservation. Certainly, the work of more than a century of science-based advocacy and support has assisted the conservation of ecosystems, species, and gene pools. Particularly in Africa, it has helped to minimize the rate of loss of species during the twentieth century. Yet the outcomes of science-based conservation have been too limited. They have not met growing demands for individual skills and competencies, or for increased institutional capacity and authority. Researchers have also done too little to help mobilize new financial investments for conservation capacity. Increased training, education, institutional development, and funding are needed to ensure that ecological and behavioral knowledge generated by research are integrated into policy and decision-making processes, to mitigate loss of habitats and range-space, whether at site or country levels.

The problems are particularly severe with respect to conservation of the great apes. Like many other large mammals, these species reproduce slowly and require extensive territories. They face significant threats to their survival in the new millennium from growing human populations, competition for land, and regimes of resource use that fragment their habitats and ranges. Success in maintaining wild populations of threatened and endangered species will therefore require us to learn, in other words innovate culturally, to significantly expand our visions, imagination, willingness, and talents. We need to find new politically, economically, and socially justifiable ways to integrate the ecological and behavioral needs of targeted species into large land/seascape area governance. Like the great apes themselves, we must be inventive.

Research–conservation partnerships dedicated to engaging context-specific political, economic, and social processes to establish and sustain the necessary conditions for conserving habitats and populations of targeted species have a potential niche in international environmental affairs during the first half of the twenty-first century. Great apes, including chimpanzees and other "charismatic" species, may serve as pilots in exploring and expanding that niche. This chapter suggests some ways forward.

International cooperation in science, conservation, and the environment

Nature conservation and pollution are the twin pillars on which the modern international environmental movement has been built (Paehlke, 1989), with conservation (including nature conservation) being the first major topic to dominate UN-based environmental debates. Even before the establishment of the UN, however, scientists were contributing to international discussions of environmental issues through organizations such as ICSU (International Council of Sciences, founded in 1931). Ecological scientists and conservationists played an important role in urging the UN to pursue nature protection issues soon after the multi-lateral body and its specialized agencies like UNESCO (United Nations Educational, Scientific and Cultural Organization) were set up in the mid-to-late 1940s.

Environmental historian John McCormick (1995) attributes UNESCO's commitment to nature conservation to the singular efforts of its first Director General, Julian Huxley. McCormick notes that "the word 'conservation' appeared in UNESCO's constitution, but in relation to books and works of art, not natural resources." Yet in 1947 Julian Huxley convinced the UNESCO General Conference that the "enjoyment of nature was part of culture, and that preservation of rare and interesting animals and plants was a scientific duty." It was Huxley's commitment that convinced UNESCO to convene the first ever technical conference on nature protection and international cooperation in ecological research: the UN Scientific Conference on the Conservation and Utilization of Resources (UNSCCUR) held in Lake Success, New York, USA in 1949.

In the 1960s and the 1970s many features of the current global environmental movement began to emerge. Awareness of pollution and its impacts were heightened among the public of developed countries, and led scholars to question the paradigm that underpinned "development" of those industrialized nations. A political "South" consisting largely of non- or less-industrialized nations was coalescing around the conviction that environment and development are two sides of the same coin. The "South" was suspicious of the environmental agenda of the political "North," which appeared to advocate the search for alternatives to tried and tested ways of building national economic prosperity. But together with others, Maurice Strong, the Secretary General of the 1972 Stockholm Conference on the Human Environment, persuaded the "North" and "South" to convene in the newly established UNEP (United Nations Environment Program) to discuss interlinked problems of the environment and development in a regular and continuing dialog.

In UNESCO two streams of thought emerged in the late 1960s and led to the articulation in the early 1970s of initiatives that were in line with global trends of the day. They placed UNESCO as an organization promoting research–conservation partnerships at the interface of environmental conservation and development.

- First, in 1969 the Biosphere Conference continued the theme of international cooperation in ecological research that had been first explored in UNSCCUR in 1949. This led to the launch, in 1971, of the Man and the Biosphere (MAB) Program.
- Second, in 1972 the sixteenth session of the General Conference of UNESCO adopted the "World Heritage Convention," entitled "The Convention Concerning the Protection of the World's Cultural and Natural Heritage."

MAB AND WORLD HERITAGE

History and significance

The MAB Program and the World Heritage Convention, together with the 1971 Ramsar Convention on Wetlands, provide the frameworks for the formal designation of sites where international environmental policy can be linked to national and local contexts, agendas, and action. All of them are products of the early phases of the "new environmental age" (Nicholson, 1987) that dawned in the 1970s. MAB and the World Heritage Convention continue to be coordinated by UNESCO in Paris, France. The Secretariat of the Ramsar Convention is located in the premises of IUCN in Gland, Switzerland, but UNESCO continues to be the UN body that receives the instruments of ratification, adhesion, or acceptance from countries wishing to become Party to that Convention. Discussions in this chapter are limited to my experience since 1986 in working with the MAB Program and the World Heritage Convention.

A prime function of the MAB Program is to designate UNESCO Biosphere Reserves. Currently, the World Network of Biosphere Reserves counts 529 sites in 105 countries. By comparison, the World Heritage List of UNESCO now includes 851 sites, of which 166 are natural and 25 are mixed (natural/cultural) sites, dedicated to conserving outstanding biodiversity, ecosystem, earth-system, and aesthetic values. The rest of the 666 World Heritage sites are recognized predominantly for their cultural value but often comprise landscapes and other entries of regional, national, and local biodiversity significance. Detailed accounts of the history of the MAB Program and biosphere reserves are given by Batisse (1982, 1986 and 1993). Ishwaran (2004) provides a recent description of the evolution of the World Heritage Convention and its operations, particularly with regard to natural and mixed sites.

In order to be recognized as a World Heritage Site based on biodiversity, ecosystem, and associated values, a nominated area must be legally protected. Sites proposed as biosphere reserves, in contrast, must contain substantial areas outside of legally protected core zones, i.e., in buffer and transition areas that have resident human populations. Biosphere reserves are sites for experimenting with the MAB approach of taking conservation "out-of-the-box," and integrating it into broader ecosystem or landscape level planning with the active participation of interdisciplinary research teams and concerned stakeholders in governance and management of the area. Of the total surface area of the 187 biosphere reserves included in UNESCO's World Network since 1995, only

11% of the designated area is legally protected, generally as a core surrounded by buffer zones and transition areas; the rest of the 89% of the total area is in buffer zones (32%) and transition areas (57%), respectively. Buffer and transition areas of biosphere reserves, however, are spaces where the environmental, economic, and social dimensions of development must be given balanced consideration in policy and decision making. The goal of the biosphere reserve designation is to demonstrate the feasibility of positive and mutually reinforcing conservation and development linkages at the landscape level, aided by research and monitoring, education, training and institution building, and participation of local communities in governance and management.

Nominations for World Heritage site status based on biodiversity criteria are also required to identify a buffer zone, but in the majority of the cases the buffer zone is not included in the area to be accorded World Heritage status. Instead, the buffer zone is a surrounding area that is integral to the management of the designated area (the World Heritage Site) because it affords protection and safeguarding of the outstanding values of the site itself, deemed to be humanity's common heritage.

Relationship to research

World Heritage sites and biosphere reserves are preferred places for Member States (or Parties) to meet their international environmental and development commitments, which they can do with support from UNESCO and its partners. For example, various commitments are required by the Convention on Biological Diversity (CBD), which was adopted at the Rio de Janeiro (Brazil) Summit on Environment and Development in 1992. One effect of the CBD is that, in general, the issue of biodiversity conservation has been shifted away from the realm of science and biological research, and more towards institutional development policy (Medley and Kalibo, 2007). Nevertheless, research is still important, and World Heritage sites and biosphere reserves are ideal settings for it. Thus, in order to assess progress in the global commitment made by Parties to the CBD in 2002, the outcomes of national and local actions to reduce biodiversity loss must be measured and verified by the year 2010. In an attempt to develop a practical index of success in meeting the "CBD-2010 target" to reduce biodiversity loss significantly, Scholes and Biggs (2005) have proposed a biodiversity intactness index (BII) that depends to a considerable extent on the status of biodiversity outside of legally Protected Areas. It is therefore helpful to take advantage of areas like the Mata Atlantic Biosphere Reserve of Northeastern Brazil. Mata extends over

29 473 484 hectares of which only 4 052 544 hectares are in legally protected core zones, including three World Heritage sites: Iguacu National Park (170 000 hectares), Atlantic Forest South-East Reserves (468 193 hectares) and the Discovery Coast Atlantic Forest Reserves (111 930 hectares). More than 25 000 000 hectares of the Mata Atlantica Biosphere Reserve are public lands included in buffer and transition zones where communities live and use land and resources to improve their socio-economic well-being and where policy and action to integrate biodiversity conservation and socioeconomic development are on-going processes. Such examples are encouraging UNESCO to initiate new research in cooperation with Parties to the CBD to assess their performance and prospects for the future through the use of BII or similar indicators. Land- or seascapes that have twin World Heritage and biosphere reserve status, such as Mata Atlantica, may suit such research efforts to be promoted by UNESCO.

Research and conservation collaboration, on a regional as well as thematic basis, is also encouraged via networks among World Heritage sites and biosphere reserves. In addition to enabling countries to meet their commitments and obligations to multilateral environmental treaties such as the CBD, these networks facilitate international cooperation with regard to environment and development policies. To this end, many World Heritage sites and biosphere reserves are part of other international and regional (e.g., European) long-term ecological and socioecological research networks and partnerships.

Although scientific research therefore has an important role to play in these conservation areas, there are new and difficult challenges ahead. Research–conservation partnerships of the future must engage increasingly in the political, economic and social processes that are required to conserve habitats and populations of endangered and threatened species at the land and seascape level. The problem is particularly important for biosphere reserves where more than 80%–90% of the designated area is outside of legally protected cores, because in these circumstances the ability of research–conservation partnerships to influence decision-making for the whole of the biosphere reserve can become a straightforward issue of credibility.

Consider the Amboseli Biosphere Reserve of Kenya, for example. Although this reserve covers 5500 km^2, its legally protected core (the Amboseli National Park) is only 392 km^2. Recent analyses undertaken jointly by the Amboseli Elephant Trust, MAB National Committee of Kenya and UNESCO highlighted the desirability of setting aside additional areas for biodiversity conservation across buffer zone and transition areas in and outside the Amboseli Biosphere Reserve, in order to achieve a key goal of linking the Amboseli ecosystem (in Kenya) directly with the Mount Kilimanjaro World Heritage site (in Tanzania) (Croze *et al.*, 2006). While this recommendation is based on valid research, the processes and incentives that will convince governance and management regimes in the Amboseli–Kilimanjaro trans-border ecosystem complex to take up and implement these recommendations are not in place. So, the future of the Amboseli–Kilimanjaro ecosystem complex depends not merely on good science, but also on changes at local, national and global levels of governance. Such changes will require stakeholders to commit to a long-term research–conservation partnership and engage in political, economic and social processes. The 21st Century Vision and Action Plan for the Ecological Society of America (2004) has expressed the need and the challenge succinctly: "To accomplish this, ecologists will have to forge partnerships in scales and in forms that they have not traditionally used. These alliances must implement action plans within three visionary areas: enhance the extent to which the decisions are ecologically informed; advance innovative ecological research directed at sustainability of an over-crowded planet; and stimulate cultural changes within the science itself that build a forward-looking international ecology." The need for scientists to commit more of their time to conservation politics is thus a worldwide phenomenon.

UNEP–UNESCO GREAT APE SURVIVAL PROJECT (GRASP)

UNESCO's relationship to GRASP

Partnerships that bring together individuals and institutions representing the public and private sector and the civil society, to commit themselves to long-term engagement in environmental, biodiversity, and conservation issues, are increasingly the norm in intergovernmental affairs too. The prime UNESCO example with respect to great ape conservation is our commitment to the Great Ape Survival Project (GRASP). GRASP is a United Nations Type II partnership, bringing together UNESCO and UNEP, several leading international and national non-governmental organizations, the great ape range states, other countries interested in and committed to great ape conservation, and a small number of private sector interests. The aim of this partnership is to conserve chimpanzees, bonobos, gorillas, and orangutans.

UNESCO brings to the partnership several key assets. First, UNESCO coordinates regional, subregional, and thematic research and conservation networks in various World Heritage sites and biosphere reserves in

Africa and Asia where great apes are found. Second, under the MAB Program UNESCO supports research and capacity building in integrated approaches to the management of lands in the humid and subhumid tropics. The MAB National Committees of Brazil and China, for example, have provided forums and instruments for academic and research communities to interact with policy professionals to influence decision making on conservation–development relationships at local, regional and national levels. Third, UNESCO promotes links from conservation areas to educational systems and networks such as the Associated Schools Network of UNESCO, which has nearly 7000 participating schools. Fourth, UNESCO can mobilize partners and resources to address and solve context-specific problems to enhance prospects for the conservation of great apes' populations and habitats.

For example, since UNESCO joined GRASP at the World Summit on Sustainable Development in 2002, it has mobilized financial resources to support 15–20 African young scientists to undertake great apes research and for organizing great apes exhibitions in East, West and Central Africa and Southeast Asia. UNESCO has had a long and close relationship with the Government of the Democratic Republic of the Congo (DRC) with regard to World Heritage and MAB activities: UNESCO coordinated the UN Foundation and Belgian Government financed projects for conserving World Heritage sites of Salonga, Virunga, Kahuzi-Biega and Garamba National Parks, and the Okapi Wildlife Sanctuary during times of war and conflict (1999–2005); and the work of the Kinshasa-based, EU-funded Regional School for Postgraduate Training in Integrated Approaches to Land Management in the Tropics (the French acronym being ERAIFT) is coordinated by the MAB Secretariat. These relationships enabled UNESCO and UNEP to convene the first intergovernmental conference on GRASP in Kinshasa in 2005 that resulted in the "Kinshasa Declaration" that publicly committed all GRASP partners to the conservation of the habitats and populations of great apes.

World Heritage sites for great apes?

Issues and problems of integrating biodiversity and development outside of legally Protected Areas that MAB faces in many of the biosphere reserves (as exemplified in the Amboseli Biosphere Reserve of Kenya) are being encountered repeatedly in the long-term conservation and maintenance of great ape habitats and populations. For example, Ancrenaz *et al.*, (2005) have shown that, of the estimated 11 000 orangutans in Sabah, Malaysia, about 55%–60% range in forestry lands outside legally Protected Areas. These populations occupy forests that are managed for timber production using selective felling methods. Fragmented and disturbed forests in the Sabah landscape mosaic will be an inevitable part of future orangutan home range and habitats.

However, areas outside legal protection are threatened increasingly, because oil palm plantations are spreading in many parts of the island of Borneo, both in Indonesian Kalimantan and in Malaysian Sabah and Sarawak, respectively. In many places oil palm could become a preferred form of land use to forestry, because of its short-to-medium-term benefits. In Sabah, forestry lands have reached a stage when the timber crop must be allowed to regenerate and therefore timber-based economic and employment benefits are decreasing. The Sabah Forestry Department, following discussions with UNESCO and the Scientific Commission of GRASP, assigned a total area of 450 000 hectares for experimentation with a multiple land-use approach to governance and management that includes conservation of high-value habitats and critical populations of orangutan. The Forestry Department currently has the political support for the experiment from the highest levels of the Sabah government. But financial investments needed for beginning and sustaining the experiment over a sufficiently long period are not yet in place.

The problem is that the financial investment from the Sabah Forestry Department alone is inadequate. New sources of financing linked to emerging global markets in carbon trading have expressed interest; if the negotiations are successful, they will ensure partial financial flows needed for sufficiently long periods of time, i.e., 20–25 years, for tree growth and carbon capture. But the availability of such financing is tied to the Sabah Forest Department attracting additional investments to support wildlife conservation, tourism, and other land and resource use activities critical for the sustainability of the management of the whole experimental area. UNESCO and GRASP have partnered with the Sabah government to seek resources for, and sustain the feasibility of, the experiment and work towards making it a success.

Discussions on the relevance of World Heritage to great apes' conservation among GRASP partners have raised an interesting question: should great apes (and perhaps other species that are genetically or culturally close to humans) be considered humanity's common heritage and thus deserve mechanisms that would ensure global responsibility for their conservation?

As it stands, the World Heritage Convention can only be applied to sites (immovable properties) and not to species (movable properties). Yet the recent designation of a World Heritage site for the conservation of giant panda in China has raised the possibility of linking several sites of critical importance to a single species and declaring the whole cluster as World Heritage. In the case of the giant panda, executing this idea administratively and politically was feasible; most importantly, the giant panda reserves are all concentrated in the Chinese Province of Sichuan. Applying it to great apes could be more challenging. Even in the case of the bonobo (whose entire range is within the DRC), national legal protection for the considerable extent of the bonobo range that lies outside of the Salonga National Park (which is a World Heritage site) may have to be established before a World Heritage nomination can be made. Alternatively, the whole of the bonobo range may be designated as a biosphere reserve first, with the Salonga World Heritage site as its core. Governance and management regimes for buffer zones and transition areas would be needed explicitly to commit to the maintenance of viable habitats and populations of bonobos, in the planning and implementation processes of all socioeconomic development.

Designing and nominating World Heritage areas for gorillas, orangutans and chimpanzees would be an immensely more complex task as all these species have ranges that extend across international borders separating several African, and in the case of orangutans, two Asian nations. Creating trans-border World Heritage areas and/or biosphere reserves may be easier in some cases than in others. For the mountain gorilla, Virunga National Park of DRC and Bwindi Impenetrable National Park of Uganda are already World Heritage sites and the Volcanoes National Park of Rwanda is a Biosphere Reserve. These could be the starting point for a three-country dialogue to explore the best utilization of World Heritage, biosphere reserve, and other conservation area status to ensure effective management of the entire mountain gorilla range.

World Heritage species?

The most innovative application of the World Heritage idea to species conservation would be the possible acceptance by the international community of selected species as "World Heritage" (Wrangham *et al.*, 2007). Its realization is unlikely to be possible within the World Heritage Convention as it stands to date, and creating a new convention will be resisted by most nations. Internationally, there are 870 legal agreements pertaining directly or indirectly to environmental issues and more than 152 environmental treaties (Gupta, 2001) and convention fatigue may have crept into many countries and organizations! The Great Apes World

Heritage Species Project (GAWHSP) Inc. has been set up in Boston, USA to explore the further development of the idea. A good starting point could be an international program that builds on committed research–conservation partnerships dedicated to addressing the context-specific political, economic, and social problems to make people who share range space and habitats with great apes feel that those animals are essential components of their identity and well-being. When war raged between DRC, Uganda, and Rwanda, mountain gorilla populations survived and even registered a slight increase in their numbers (see pages 109–137 in UNESCO, 2005). These animals had been important to communities and people in the animal's range and had helped them earn their livelihoods through tourism during prewar years. People's memory of the value of the animal ensured its conservation even during times when the flow of economic benefits was interrupted. This outcome was made possible largely by the presence in the region of research and conservation organizations continually interacting with local communities, administrators, and private sector interests. Such long-term presence of committed researchers and conservationists in specific places is a necessary condition for sustaining the long-term future of chimpanzees, great apes, other charismatic species, or any other component of biodiversity. The moment when Kibale is celebrating its twentieth anniversary of continuous research–conservation partnership is an appropriate moment to call for the encouragement of the establishment of such partnerships in many more places.

SUMMARY

Long-term research conservation partnerships, such as the one celebrating 20 years of cooperation in the Kibale Forest of Uganda, will be critical in improving prospects for the conservation of biodiversity components in the twenty-first century, particularly in areas outside of legally Protected Areas in land/seascapes. UNESCO's biosphere reserve and World Heritage site networks are well placed for the development, promotion, and support of such partnerships. UNEP and UNESCO's involvement in GRASP provides a timely opportunity to experiment with the development of research–conservation partnerships that commit to long-term engagement with context-specific political, social, and economic processes essential to making local communities, public administrators, and private sector representatives adopt sustainable development practices that allow for viable great ape habitats and populations in the landscape.

REFERENCES

Ancrenaz, M., Gimenez, O., Ambu, L. *et al.* (2005). Aerial surveys give new estimates for orang utans in Sabah, Malaysia. *PLoS Biology*, **3**, e3.

Batisse, M. (1982). The biosphere reserve: a tool for environmental conservation and management. *Environmental Conservation*, **9**, 101–111.

Batisse, M. (1986). Developing and focusing the biosphere reserve concept. *Nature and Resources*, **XXII**, 1–11.

Batisse, M. (1993). The silver jubilee of MAB and its revival. *Environmental Conservation*, **20**, 107–112.

Croze, H., Sayialel, S., and Sitonik, D. (2006). What's on in the ecosystem? Amboseli as a biosphere reserve. A compendium of conservation and management activities in the Amboseli ecosystem. Amboseli Elephant Trust, Nairobi, Kenya.

Ecological Society of America (2004). 21st century vision and action plan for the Ecological Society of America. Report of the Ecological Visions Committee of the Ecological Society of America. Washington, DC, USA.

Fernandez-Armesto, F. (2004). *So You Think You're Human?* Oxford: Oxford University Press.

Gupta, J. (2001). Legitimacy in the real world: a case study of the developing countries, non-governmental organizations and climate change. In *The Legitimacy of International Organizations*, ed. J-M Coicaud and V. Heiskanen. Tokyo, New York, Paris: United Nations University Press, pp. 482–518.

Ishwaran, N. (2004). International conservation diplomacy and the World Heritage Convention. *Journal of International Wildlife Law and Policy*, **7**, 43–56.

McCormick, J. (1995). *The Global Environmental Movement*. Hoboken: John Wiley & Sons.

Medley, K. E. and Kalibo, H. W. (2007). Global localism: recentering the research agenda for biodiversity conservation. *Natural Resources Forum*, **31**, 151–161.

Nicholson, M. (1987). *The New Environmental Age*. Cambridge: Cambridge University Press.

Paehlke, R. C. (1989). *Environmentalism and the Future of Progressive Politics*. New Haven: Yale University Press.

Scholes, R. J. and Biggs, R. (2005). A biodiversity intactness index. *Nature*, **434**, 45–49.

UNESCO. (2005). Promoting and preserving Congolese Heritage: linking biological and cultural diversity. *Proceedings of the Conference and Workshops, UNESCO*, 13–17 September 2004. World Heritage Papers, 17.

Wrangham, R. W., Hagel, G., Leighton, M., Marshall, A., Waldau, P., and Nishida, T. (2007). The Great Ape World Heritage Species Project. In *Conservation in the 21st Century: Gorillas as a Case Study*, ed. T. Stoinski, D. Steklis, and P. Mehlman. New York: Springer, pp. 282–295.

RICHARD WRANGHAM AND ELIZABETH ROSS

20

Long-term research and conservation: the way forward

This book has described the conservation activities generated by seven African long-term research stations. They are the sites where research has been on-going for the longest period – the six oldest sites for chimpanzees, and the single oldest site for gorillas. In all of them the impacts on conservation appear to be diverse and positive, suggesting that the presence of researchers is an important predictor of conservation success. There is also informal evidence suggesting that the conservation status of these sites is better than comparable areas without such research. Gombe National Park, for example, was once surrounded by forest but is now an island of forest in a sea of agriculture. However, our sample of field stations is small, and various kinds of bias could temper the conclusion that long-term research has been responsible for conservation. In this chapter, therefore, we first summarize the conservation activities carried out by long-term researchers. We then suggest that the increased establishment and support of field stations might lead to improved conservation success in the future.

THE CONSERVATION IMPACTS OF LONG-TERM RESEARCH

Conservation consequences emanating from long-term research are diverse both within and across sites. Within sites, the variety is illustrated by the case considered in the greatest detail, Makerere University Biological Field Station (MUBFS) at Kibale National Park in Uganda. Across sites, researchers from the six other sites considered in this book (Bossou, Budongo, Gombe, Mahale, Taï, Virungas) report on major activities that vary from ecotourism and education to community-based conservation projects. The activities from these seven sites fall into two broad groups reflecting shorter-term in-habitat activities, or longer-term building of support for conservation.

First, in-habitat activities by researchers include obtaining data (Mapesa, Chapter 2; Olupot and Plumptre, Chapter 3), introducing new techniques (Laporte *et al.*, Chapter 5), providing and testing hypotheses (Chapman *et al.*, Chapter 6; Lwanga and Basuta, Chapter 7; Goldberg *et al.*, Chapter 8; Babweteera *et al.*, Chapter 13), proposing intervention programs (Chapman *et al.*, Chapter 6, Goldberg *et al.*, Chapter 8), and implementing conservation, for example, by forest restoration (Matsuzawa and Kourouma, Chapter 17), snare removal programs (Babweteera *et al.*, Chapter 13) or community programs (Collins and Goodall, Chapter 14). In addition, the role of researchers in monitoring their study populations for demographic changes, particularly caused by disease, is becoming increasingly important (Goldberg *et al.*, Chapter 8; Nishida and Nakamura, Chapter 15; Boesch *et al.*, Chapter 16). These contributions are valuable for managers of protected areas, because, as Struhsaker notes in Chapter 4, every Protected Area has its own particular set of problems and species, so that conservation solutions need to be designed locally on the basis of specific data sets. Protected Area Authorities rarely have adequate personnel and funds to achieve all their goals, so the contributions from research stations can be welcome (Mapesa, Chapter 2). As Olupot and Plumptre explain in Chapter 4, for this reason alone "pure" researchers have much to offer.

Second, long-term contributions to support for conservation include training, promoting national and international awareness, contributing to the development of economic benefits, and community outreach. In his role as Director of the Uganda Wildlife Authority (UWA), Mapesa (Chapter 2) draws attention to the role of research stations in building technical capacity. For example, stations such as MUBFS provide opportunities for tertiary education for both national and foreign students. Trevelyan and Nuttman (Chapter 9) report on the use of MUBFS for field courses in tropical biology. Student experiences seem likely both to promote interest in the inherent value of tropical forest habitats and to foster skills and activities relevant to their conservation.

In several sites the role of research stations in education appears to be important for preserving biodiversity, with target audiences ranging from local populations to government agencies and the wider international community. Boesch *et al.* (Chapter 16) describe an unusual theatrical effort to sensitize villagers around Taï National Park, while Babweteera *et al.* (Chapter 13) and Williamson and Fawcett (Chapter 18) note that researchers help build constituencies of interest that lead to political support. More generally, research stations create groups of local people who care about the future of particular species and habitats. For example,

Williamson and Fawcett (Chapter 18) note that the dedication of gorilla trackers has been important in maintaining the research project, with all its spin-offs. The equivalent could be said for fieldworkers in all the sites. The ultimate education efforts are those that help persuade governments to raise the protection status of areas, as appears to have happened in several of the sites considered here (e.g., Nishida and Nakamura, Chapter 15).

The opportunity for researchers to help to develop economic benefits varies with different protected areas. Great apes lend themselves to ecotourism projects, several of which have been stimulated by research activities (Mugisha, Chapter 11; Nishida and Nakamura, Chapter 15; Williamson and Fawcett, Chapter 18). As Nishida and Nakamura note, in particular, ecotourism can easily slide into exploitative tourism, so that the continuing attention of researchers working collaboratively with a Protected Area Authority is potentially valuable.

Several chapters draw attention to the merits of integrating biological research with social science, so that the impacts of conservation actions on the local community can be understood better (e. g., Lwanga and Basuta, Chapter 7; Goldman *et al.*, Chapter 12). Kasenene and Ross (Chapter 10) and Collins and Goodall (Chapter 14) address the problem of community relations in a complementary manner by stressing the importance of long-term relationships for building trust between conservationists and local communities, enabling them to bring potentially large resources for local benefit. It seems likely that every research station involves the development of important personal and institutional relationships with high potential for generating positive action.

The significance of personal relationships and local contexts, we suggest, is a particularly valuable lesson from this book. All of the diverse conservation outcomes are more fruitful when built on trust and detailed understanding of context. For example, Mapesa (Chapter 2) and Olupot and Plumptre (Chapter 3) note that research information is vital for the design of conservation strategies implemented by Protected Area Authorities (PAAs). But, in practice, how effectively PAAs communicate their needs to researchers, and how effectively researchers assist, depend critically on personal willingness on either side. Much the same could be said for any of the initiatives described in other chapters. Long-term involvement builds trust, and trust facilitates cooperation.

LONG-TERM RESEARCH AND THE FUTURE OF CONSERVATION

Much conservation activity initiated by the developed world in tropical forests stems from the major conservation organizations, such as the

Wildlife Conservation Society (WCS), the Worldwide Fund for Nature (WWF), Conservation International (CI), and the African Wildlife Foundation (AWF). It is supported also by UNESCO and UNEP (Redmond and Virtue, Foreword), and by many different funding sources such as foreign aid programs and the Global Environmental Facility (GEF). Many of these sources have a substantial history of supporting the kinds of long-term research that we have highlighted in this book. They have made possible the concept that a particularly effective conservation strategy for a vulnerable habitat is to encourage research, leading to the establishment of a research station.

This book makes the case that long-term research stations are associated regularly with a wide range of conservation activities, many of which apparently happen partly because a research station is present – making it easy for new initiatives to be developed, for long-term data to be used, and for personal relationships that foster creative plans to flourish. Admittedly, we cannot yet decide for certain how much of the conservation activity can be attributed to the existence of the research station, or how effective the activity is for conserving populations and habitats. But the obvious explanation is that habitats in which a long-term program of research is established are more likely to be protected as a result. The specific impacts on public awareness, government involvement, protection status, monitoring, education, community outreach, provision of research data, conservation interventions, and development of economic benefits will vary from site to site. The common theme suggested by the experiences described in this book is that, although the benefits vary, they will be important.

We therefore hope this book will lead to increased consideration of the potential benefits of nurturing short-term research studies into long-term programs. If evidence continues to mount that one of the best ways to conserve habitat is for it to host a research station, there are obvious implications for conservationists and their political and financial backers. The next step, therefore, is to design studies that evaluate the impact of research stations on conservation as systematically as possible. But, even now, a reasonable global conservation strategy for tropical forests would include the aim of establishing a research station in every major forest. The hope that long-term field studies can make a substantial difference to conservation should surely help motivate future generations of researchers.

Index